Cyberwar, Netwar and the Revolution in Military Affairs

Also by Edward Halpin

HUMAN RIGHTS AND THE INTERNET (*co-editor with S. Hick and S. Hoskins,* 2000)

Cyberwar, Netwar and the Revolution in Military Affairs

Edited by

Dr Edward Halpin,
Leeds Metropolitan University, UK

Dr Philippa Trevorrow,
Leeds Metropolitan University, UK

Professor David Webb
Leeds Metropolitan University, UK

and

Dr Steve Wright
Leeds Metropolitan University, UK

First published 2006 by
PALGRAVE MACMILLAN
Houndmills, Basingstoke, Hampshire RG21 6XS and
175 Fifth Avenue, New York, N.Y. 10010
Companies and representatives throughout the world

PALGRAVE MACMILLAN is the global academic imprint of the Palgrave Macmillan division of St. Martin's Press, LLC and of Palgrave Macmillan Ltd. Macmillan® is a registered trademark in the United States, United Kingdom and other countries. Palgrave is a registered trademark in the European Union and other countries.

ISBN-13: 978–1–4039–8717–4 hardback

This book is printed on paper suitable for recycling and made from fully managed and sustained forest sources.

A catalogue record for this book is available from the British Library.

Library of Congress Cataloging-in-Publication Data
Cyberwar, netwar, and the revolution in military affairs / edited by Edward Halpin... [et al.].
   p.   cm.
Includes bibliographical references and index.
ISBN 978-1-4039-8717-4 (cloth)
1. Information warfare.   2. Military art and science—History—21st century.   I. Halpin, Edward F.
U163.C938 2006
355.3'43—dc22   2005045613

10   9   8   7   6   5   4   3   2   1
15   14   13   12   11   10   09   08   07   06

Contents

Part II: Implications of the Problem

List of Tables

Preface

Gary Chapman, Diego Latella and Professor Carlo Schaerf

The subject of this volume relates to the work of the International School on Disarmament and Research on Conflicts (ISODARCO).

There is no unique definition for such things as Information Warfare (IW). For instance, according to the US Department of Defense (DoD), it could be defined as: those actions taken for achieving information superiority by means of acting on the information of the adversary, on their information-based processes, on their information systems and on their computer networks, while at the same time levering on and defending one's own information. In a more explicit way, the Ministry of Defence in the UK defines Information Warfare as the deliberate and systematic attack on critical information activities, which aims at exploiting, modifying and compromising information and at interrupting services. Despite the lack of unique definitions for IW/CW, NW and RMA, a lively discussion on their nature, the threats they bring, the possible countermeasures to be undertaken by nation-states, as well as other organizations, is taking place in political and military circles as well as academia. Issues of major importance in such discussions are: the relation between computers and regional defence; the threat of 'cyber terrorism' as well as 'cyberwar'; new forms of group organization like 'networks' and how information technology supports them; the impact of information technology developments on military doctrine and organization of military forces.

Without any doubt, some of the above issues are connected to real threats, but the dimension of such threats is far from being fully assessed and understood. Thus, as often happens when new scenarios are elaborated, a proliferation of myths related to CW, NW and RMA are also taking place and they bring with them a possible real threat of widespread global surveillance. Ironically enough, information technology is itself supporting the proliferation of such myths by providing a 'virtual space' where a great deal of the ongoing discussion on the above issues is taking place.

ISODARCO work in this area has been made possible by the generous contributions of several Foundations and Institutions. It is a pleasure to thank: The John D. and Catherine MacArthur Foundation, and, in particular, Dr Kennette Benedict; The Fondazione Opera Campana dei Caduti, Rovereto, and in particular its' Reggente Dr Pietro Monti; the Giunta della Provincia Autonoma di Trento; the Università degli Studi di Roma 'Tor Vergata', Dipartimento di Fisica; the Istituto Trentino di Cultura-irst, Trento; the Università degli Studi di Trento; and the CNR Istituto di Scienza e Tecnologie dell'Informazione 'A. Faedo', Pisa.

It is a pleasure to also thank Dr Mirco Elena and the administration and staff of the Istituto Salesiano 'Maria Ausiliatrice' of Trento.

All opinions expressed in this book are of a purely personal nature and do not necessarily represent the official view of either the organisers of ISODARCO the organizations with which the writers may be affiliates.

Notes on the Contributors

Martin Bayer studied War Studies at King's College London (2000–2003) with the main focus on European Defence, simulations, media, RMA, and civil–military relations. In 2004, he was awarded a Master of Peace and Security Studies from the Institute of Peace and Security Studies at the University of Hamburg (IFSH). He has given several presentations at international conferences, primarily on the depiction of violence and war, military strategy, and network-centric warfare (NCW). Currently, he is German correspondent for *Jane's Defence Weekly*.

Gary Chapman was educated at Occidental College in Los Angeles, California, and at Stanford University. He has also taught at both institutions. Currently, he is director of 'The 21st Century Project' at the Lyndon B. Johnson School of Public Affairs, the graduate school of public policy at the University of Texas in Austin. He is associate director of the Telecommunications and Information Policy Institute, also at the University of Texas. Chapman currently writes a bi-weekly column for *The Austin American-Statesman*, a column which runs every other Friday on the op-ed page of the newspaper.

Geoffrey Darnton is Head of Knowledge Transfer for the Institute of Business and Law, Bournemouth University and Programme Leader for their MA in Information Systems Management. He is a Chartered Engineer, Chartered Statistician, and Certified Management Consultant. He compiled, edited and wrote much of the London Nuclear Warfare Tribunal summary and judgment. His written publications include *Information in the Enterprise* and *Business Process Analysis*. He has contributed to other books, including *Information Systems in Organizations*. Information Warfare is one of his main research interests.

Dr Stefan Fritsch is currently a Lecturer for International Relations in the Department of History and Political Science at the University of Salzburg. His research interests include technology and International relations/International political economy, multinational corporations, issues of globalization and global governance, and European integration/EU.

Dr Edward Halpin is Director of the Praxis Centre at Leeds Metropolitan University, UK. He has researched human and child rights for many years, including work for the European Parliament Scientific and Technical Options Assessment (STOA) Unit. He co-edited the book *Human Rights and* *the Internet* (Palgrave Macmillan) and has published many articles in this subject area. In addition to working on social informatics within the School of Information Management, he is involved in teaching peace and conflict resolution in the School of Applied Global Ethics at Leeds Metropolitan University.

Dr Gus Hosein is a Visiting Fellow in the Department of Information Systems at the London School of Economics and Political Science. He is also a Senior Fellow with Privacy International where he directs the Terrorism in the Open Society programme. He also co-ordinates the Policy Laundering Project, a joint initiative between PI, Statewatch, and the American Civil Liberties Union. He has a PhD from the University of London and a B.Math(Hons) from the University of Waterloo.

Dr Bruce Larkin studied at the University of Chicago (1950–56) and Harvard University (1960–65). He then joined the founding faculty of the University of California at Santa Cruz, where he is now Professor Emeritus of Politics. His books include *Nuclear Designs: Great Britain, France, and China in the Global Governance of Nuclear Arms* (1996) and a study in war theory, *War Stories* (2001).

Diego Latella studied Computer Science at the University of Pisa where he graduated (with honours) in 1983. In 1986 he joined CNUCE, an Institute of the Italian National Research Council (CNR) where he currently works in the Formal Methods and Tools group. He lectures at the Department of Computer Science of the University of Pisa. He is a member of the Scientific Council of the Interdepartmental Center for Sciences for Peace (CISP) of the University of Pisa and of the Advisory Board of Privacy International. His current main research interests fall in the area of formal specification and verification models for concurrent systems and their application.

Massimo Mauro is principal administrator at the Council of the European Union, Directorate-General C 1 (Internal Market), where he is responsible for all technical harmonization legislation. He holds a dottore in fisica degree from the State University of Catania, Italy, and speaks 10 languages. In addition to his EU work, he has also held public and private sector positions with research and technology firms in the UK, Italy, Germany, and Finland. He is a full member of the Association for Computing Machinery, of Computer Professionals for Social Responsibility, and of AICA, the Italian computer society.

Mike Moore is contributing editor of the *The Bulletin of the Atomic Scientists*, a peace-and-security magazine founded at the University of Chicago in 1945 by key members of the Second World War atom bomb project (www.thebulletin.org). Previously, he was editor of the *Bulletin*; editor of *Quill* (the magazine of the Society of Professional Journalists); and an editor or reporter at the *Kansas City Star*, the *Chicago Daily News*, the *Chicago Tribune*, and the *Milwaukee Journal*.

Dr Peter G. Neumann has doctorates from Harvard and Darmstadt. He works in SRI's Computer Science Lab and is concerned with computer systems and networks, security, reliability, survivability, safety, and many risk-related issues such as voting-system integrity, crypto policy, social implications, and human needs, including privacy. He is the 2002 recipient of the National Computer System Security Award.

Fanourios Pantelogiannis is a PhD student at the Free University of Brussels (ULB). His first research work, *Russian Military Reform*, was published online by the Institute of Advanced European and International Studies. His more recent publications include *European Entrepreneur's Handbook*, published in Greek in 2004. His main research interests are the EU political system, the Russian military and international negotiations.

Lieutenant Colonel Jari Rantapelkonen graduated from the Finnish Military Academy. His current assignment is as an Information Operations Instructor for the National Defence College in Helsinki, Finland. He has published widely, writing books and articles on the philosophy of military science, operations and tactics. Currently he is preparing a PhD for the University of Helsinki and National Defence College on the phenomena of Information Warfare.

Professor Carlo Schaerf is Professor of Physics at the University of Rome 'Tor Vergata'. From 1978 to 1984 he was President of the National Commission for Nuclear Physics (INFN). Schaerf then became President of the National Committee for the Physical Sciences at the Ministry of Public Education (1984–1987). In 1966 Professor Carlo Schaerf founded (with Professor Edoardo Amaldi) the International School on Disarmament and Research on Conflicts (ISODARCO). Professor Schaerf became its Director in 1970 and Director and Chairman of the Board in 1991. He is editor of some 25 books on disarmament and physics, as well as the author of approximately one hundred articles in international refereed scientific journals.

Gian Piero Siroli has a BSc, a Diploma and a 'Corso di Perfezionamento' in Physics from the University of Bologna. He is currently lecturing on 'Computer and Network Security' in the Department of Physics at the University of Bologna and on 'Information Technologies and International Security' in the Computer Science Department at the University of Pisa. In 1999 he participated at the UN meeting on 'Developments in the field of information and telecommunications in the context of International security', hosted by DDA and UNIDIR at Palais des Nations in Geneva.

Dr Philippa Trevorrow has a BSc B(Ed.) Hons in Mathematics and Education, and a PhD from the University of Exeter. She currently works as a Research Officer in the School of Information Management at Leeds Metropolitan University, UK. She has been involved with work in the Praxis Centre for the last two years, including issues on peace and conflict resolution, youth citizenship and E-government.

Professor David Webb is Professor of Engineering, head of the Centre for Applied Research in Engineering, and Director of the Praxis Centre at Leeds Metropolitan University. He obtained a D.Phil. in Space Physics in 1975 from the University of York and after periods as a post-doc at Bell Labs and the UK, joined the Directorate of Scientific and Technical Intelligence at the MoD in London in 1978. He moved to the Computer Unit at Leeds in 1979 and then into the School of Engineering in the early 1980s. He has published widely on the application of engineering modelling and on Nuclear Disarmament and the Militarisation of Space. He is currently working with colleagues in the Praxis Centre on the 'Study of Information and Technology in Peace, Conflict Resolution and Human Rights'.

Chris Wu (original name Wu Fan) is a native of Wenchang County, Hainan Island. He graduated from the Anhui Normal University in 1965, having studied and worked in the Department of Physics. In 1980 he relocated to the United States to study for a Masters degree in the Department of Electronics and Computer Engineering, California State University, Los Angeles. In 1987 he became a Senior Electronics Engineer at Xerox Corp's microelectronic central laboratory. He is currently Chief Editor of *China Affairs* and a Research Fellow at the Centre for Modern China. His research interests include: politics, economy, military affairs, diplomatic affairs, and the relationship between Mainland China and Taiwan.

Dr Steve Wright is a Visiting Professor at Leeds Metropolitan University, UK, Chair of Privacy International and the former Director of the Omega Foundation. His recent EU research tracked the armourers of the torturers. He is best known for his European Parliamentary report highlighting the US global telecommunications interception network – Echelon.

Glossary

ABL	Air-Borne Laser
ABM	Anti-Ballistic Missile
ALMV	Air-Launched Miniature Vehicle
AMHS	Automated Message Handling System
APS	American Physical Society
ARP	Address Resolution Protocol
ASAT	Anti-Satellite
ASEM	Asia–Europe Meeting
ASW	Anti-Submarine Weapon
ATM	Asynchronous Transfer Mode
BINAC	Binary Automatic Computer
BMC3	Battle Management, Command, Control and Communication
BMD	Ballistic Missile Defence
C3I	Command, Control, Communications and Intelligence
C4I	Command, Control, Communications, Computers and Intelligence
C4ISR	Command, Control, Communications, Computing, Intelligence, Surveillance and Reconnaissance
CAVE	Computer-Aided Virtual Environment
cbICT	Computer-Based Information and Communications Technology
CDI	Center for Defense Information
CERT	Computer Emergency Response Team
CIAO	Critical Infrastructure Assurance Centre
CIMIC	Civil–Military Cooperation
CIS	Commonwealth of Independent States
CMR	Civil–Military Relations
COE	Common Operating Environment
CoE	Council of Europe
COP	Common Operational Picture
COTS	Commercial Off-The-Shelf
CPU	Central Processing Unit
CSRS	Counter Surveillance Reconnaissance System
CW	Cyberwar
DARPA	Defence Advanced Research Projects Agency
DDA	Department of Disarmament Affairs
DDoS	Distributed Denial of Service
DG	Directorate General
DISN	Defence Information System Network

DNS	Domain Name System
DoD	Department of Defense
DPS	Defence Support Program
DSCS	Defence Satellite Communications System
DSWA	Defence Special Weapons Agency
ECHR	European Court of Human Rights
EDSAC	Electronic Delay Storage Automatic Computer
ELINT	Electronic Intelligence
EMP	Electromagnetic Pulse
ENIAC	Electronic Numerical Integrator and Calculator
ESA	European Space Agency
EW	Electronic Warfare
FIDNet	Federal Intrusion Detection Network
G8	Group of Eight industrialized countries
GATT	General Agreement on Tariffs and Trade
GBR	Ground Based Radar
GCCS	Global Command and Control System
GCCS-T	Global Command and Control System Top Secret
Glonass	Global National Satellite System
GPS	Global Positioning System
HPM	High-Power Microwave
ICAO	International Civil Aviation Organization
ICBM	Intercontinental Ballistic Missile
ICJ	International Court of Justice
ICT	Information and Communication Technologies
IGO	Inter-governmental organization
II	Information Infrastructure
IP	Internet Protocol
IPE	International Political Economy
IPTF	Infrastructure Protection Task Force
IR	International Relations
ISP	Internet Service Provider
IST	Information Society Technologies
IW	Information Warfare
JCS	Joint Chiefs of Staff
JDAM	Joint Direct Attack Munitions
JOPES	Joint Operation Planning and Execution System
JWID	Joint Warrior Interoperability
KE-ASAT	Kinetic Energy ASAT
KEW	Kinetic Energy Weapons
LEO	Low Earth Orbit
MDA	Missile Defence Agency
MHV	Miniature Homing Vehicle
MIC	Military-Industrial Complex

MILREP	Military Representative
MILSTAR	Military Communications Satellites
MIME	Military Industrial Media Entertainment
MIRACL	Mid-Infrared Advanced Chemical Laser
MNC	Multinational Corporations
MS	Micro Satellite
MTR	Military-Technological Revolution
NAS	National Airspace System
NATO	North Atlantic Treaty Organization
NCW	Network-Centric Warfare
NES	Network Encryption System
NGO	Nongovernmental organization
NIAC	National Infrastructure Assurance Council
NIPC	National Infrastructure Protection Centre
NIPRNET	Non-classified Internet Protocol Network
NMCC	National Military Command Centre
NMCI	Navy Marine Corps Intranet
NMD	National Missile Defence
NOC	Network Operations Centre
NSA	National Security Agency
NSF	National Science Foundation
NW	Netwar
OECD	Organization for Economic Cooperation and Development
OSI	Office of Strategic Influence
PCCIP	President's Commission on Critical Infrastructure Protection
PDD	Presidential Decision Directive
PGM	Precision Guided Munitions
PLA	People's Liberation Army
PoWs	Prisoners of War
PSTN	Public Switched Telephone Networks
R&D	Research and Development
RMA	Revolution in Military Affairs
RSTA	Reconnaissance, Surveillance and Target Acquisition
SAIC	Science Applications International Corporation
SAM	Surface-to-Air Missile
SAR	Synthetic Aperture Radar
SATCOM	Satellite Communications System
SBIRS	Space-Based Infrared System
SBL	Space-Based Lasers
SBR	Space-Based Radar
SCADA	Supervisory Control and Data Acquisition
SCP	Strategic Computing Program
SDI	Strategic Defence Initiative
SHF	Super High Frequency

SIOP	Single Integrated Operational Plan
SIPRNet	Secret Internet Protocol Router Network
SLBM	Sea-Launched Ballistic Missiles
SONET	Synchronous Optical NETworks
SRAM	Short Range Attack Missile
STA	Scientific-Technical Revolution
STOA	Scientific and Technological Options Assessment
STSS	Space Tracking and Surveillance Systems
SWPS	Strategic Warfare Planning System
TFAP	Trade Facilitation Action Plan
THAAD	Theatre High Altitude Area Defence
TRADOC	Training and Doctrine Command
T-SAT	Transformational SATCOM
UAV	Unmanned Aerial Vehicles
VPN	Virtual Private Network
WAN	Wide-Area Networks
WEP	Wired Equivalent Privacy
WMD	Weapons of Mass Destruction
WTC	World Trade Centre
WTO	World Trade Organization
WWMCCS	Worldwide Military Command and Control System

Part I

Cyberwar, Netwar and the Revolution in Military Affairs: Defining the Issues

1
Defining the Issues

Dr Philippa Trevorrow, Dr Steve Wright,
Professor David Webb and Dr Edward Halpin

The purpose of this book is to explore key emergent information technology developments for managing conflict, waging war and creating dysfunction within modern societies which are dependent on continuous information flows. It considers how the challenge is being addressed and assesses the longer-term implications and risks of these new approaches to conflict management and control. It is essentially composed of four substantive parts. Part I seeks to define the issues. Part II explores the implications of the problem and Part III presents some different (non-western) country perspectives. Finally, Part IV questions what is being done and must be done if we are to avoid being overwhelmed by competing and contradictory paradigms. The conclusion takes a tentative glimpse at innovations on the horizon and the social and political implications and ramifications.

Cyberwar, Information Warfare (IW), Netwar and the Revolution in Military Affairs (RMA) are terms that have been widely used by military observers for over a decade. In the early 1990s, in the immediate past-Cold War World, researchers such as Ronfeldt and Arquilla[1] who worked for the Rand Corporation, gave an account of what they saw as a new 'high-tech' model of warfare. The theory created by them had already gained credibility within the US military establishment by 1995,[2] though recent writers have claimed that the theoretical arguments are flawed and incomplete.[3] Much of the early discussion was either devoted to the threat to society posed by free-lance hackers, or investigated broader theoretical possibilities. Similarly, the debate on the RMA was essentially a speculation in futuristic possibilities which did not yet have a public budget line attached. All that has now changed.

Whilst the attacks of September 11 2001 have been identified as the key events in transforming individual state perspectives on these matters, such a view is largely an exaggeration. A great many other factors have also changed the military paradigms associated with the Cold War. Modern weapons are increasingly being seen as systems where the warheads and delivery mechanisms are just one component – the muscle deployed by a sophisticated and intelligent cybernetic nervous system. Real-time target acquisition data are handled through complex networks of communication, command and control systems (C3I), and war fighters recognise that it is becoming more effective to attack these information systems directly. The manner in which an efficient and effective attack can be launched on a tele-communications infrastructure may require different strategies and employ different weapons – including ones that fire electrons rather than hot pre-fragmented metal and explosives. The fact that modern societies, and their accompanying militaries, are increasingly dependent on an information infrastructure has inevitably led to a deepening analysis of what is appro-priate in exploring future threats, vulnerabilities and opportunities. As a former US Air Force Chief of Staff succinctly put it: 'dominating the information spectrum is as critical to conflict now as occupying the land or controlling the air has been in the past.'[4]

An additional factor is the extent to which a new military ideology of 'non-lethal weaponry and tactics' has begun to permeate military thinking, so that it became a formal North Atlantic Treaty Organization (NATO) commitment in 1998. Such so-called 'soft kill' approaches are always backed up with lethal force but the ideology is persuasive: why not take cities intact instead of pulverizing them into expensive rebuilding sites? So, the develop-ment of strategies and technologies which can incapacitate other technologies form an integral part of such thinking. Whilst September 11 has accelerated the pace of such logic, 'homeland security models' have also increased the pace of dedicating budget lines to specific procurement plans to fulfil what the US is now referring to as 'full spectrum dominance'.

This paradigm shift has set new levels to US military thinking and budgets on the 'information battle-space'. IW is no longer seen as an amateur Sunday league adventure in hack activism, but a highly legitimate premier division activity of the world's most dominant militaries with logistics and new directed energy weapons to match. Indeed, superficially it is hard to distinguish ideas from science fiction from today's interactive war game entertainment but there are significant differences.

Martin Bayer's exploration of 'Virtual Violence and Real War' finds that, despite the outstanding audio-visual effects of modern computer war games, they are by no means either realistic or authentic. This is contrary to a media propensity to present stories of the effective use of war games for the simula-tion of, and preparation for, actual battle. Virtual soldiers can carry huge amounts of kit, restocking and recovery is only a button push away and their simulated precision weapons always hit their targets. In reality, soldiers do not have the certainty of 'bullseye warfare' and they sit around much of the time or are engaged in repetitive, mundane tasks. These enter-taining toys can lead to misconceptions of military activity which do not carry from the play station to the battlefield. New units are not easily repaired and soldiers are legally obliged to take care of civilians, not blast prisoners of war (PoWs) or refugees at every opportunity. In the future, the key military targets will not be personnel, but the electronic nervous system used to coordinate and control their behaviour – including computer networks and the Internet. Clearly, the international community needs to be prepared to meet more professional threats to the infrastructure of the Internet, individual networks, servers and so on. The question then arises as to who the key players are in the arena of electronic conflict.

Gian Piero Siroli answers this question with a perceptive introduction to IW. The dependency on information infrastructures is emphasized with the targeting of information systems and telecommunication networks becoming an important element in the defence policies of many countries. Siroli sees the exploitation of advanced IT by the military as an important driver in the quest for new warfighting techniques. For Siroli, 'information warfare is the set of activities intended to deny, corrupt or destroy an adversary's information resources including both offensive and defensive operations'.

The second substantive part of the book examines the implications of the problem of the ubiquitous dependency on information technology and the various vulnerabilities and raises many value-laden issues.

Jari Rantapelkonen cautions us about the presentation of 'Virtuous Virtual War'. The war against terrorism is not an issue of territory, he argues; rather, it is a media creation in which virtual world technology is bestowed with a virtuous or ethical dimension. It is no coincidence that in 2001 the Rendon group was hired by the Pentagon to create a positive image of new forms of warfare. Virtuous war is equated with virtual war as presented on computer networks – essentially a Military Industrial Media Entertainment (MIME) approach which disarranges reality. 'Information bombs' do not simply destroy capacity; they can also wipe out social memories, relations and international communities. Within the RMA are advanced information networks which can misinform as well as inform, large numbers of people and can facilitate the rapid dissemination of myths and rumours which cannot be verified. The literal 24/7 coverage of the Iraq War blurred the boundaries between reality and fiction. Rather than providing useful information and analysis which can be applied to establish verisimilitude, repetition can be used to reinforce one particular set of perceptions.

Peter G. Neumann in his chapter on the 'Risks of Computer-Related Technology' draws our attention to the fact that almost everything we do nowadays is dependent on computer technology. How should we deal with this? We need reliable, secure, highly available systems. What we have in fact are networks which are highly vulnerable and have many weak links. The Internet, he reasons, is a source of tremendous benefit and yields increasing opportunities for third world development, world wide commerce, education, and information flow. And yet it can offer very little resistance to coordinated attacks since little effort has been spent on making its architecture robust. Other threats to future availability and integrity include the desire of many governments to control and regulate the Web, corporations to profit from it and a general lack of management to smooth out glitches and thwart spam and the multiplicity of pornographers, swindlers, identity thieves and snoopers who increasingly inhabit cyberspace. For Neumann, the challenge is how to retain the rich opportunities offered by the Internet, whilst restraining its risks, hazards and failures.

Neumann offers us examples of computer-based failures in defence, space, aviation, environment, telecommunications, transportation, medical systems, elections, security, privacy and law enforcement, to name but a few. If we recognise that these risks are an inherent part of new systems then we might be able to build robust systems, deal with our dependencies and prevent system failure. This could become critical when algorithmic and identity recognition systems give us access to many of the goods and services we take for granted. So far, research on robust systems has been ignored in favour of market-driven developments. This lack of vision will become critical as our defence postures are underpinned by their increasing reliance on vulnerable ICT systems.

However, despite failures, and financial and technological setbacks, many US defence programmes are continuing apace.

Dave Webb takes a look at 'Missile Defence' and asks whether this system is merely a by-product of the 'first steps towards war in space?' Webb also takes note of the increasing reliance of the military on space-based systems and examines the vulnerability of these systems to attack. It is this vulnerability that leads to the desire of the US Space Command to control space and position itself so that it can deny access of space to others when it deems necessary. The accelerating US investment in military space technology generally, and anti-satellite systems in particular, are indicators of this aspect of RMA, although, as Webb points out, the increase in traffic may be too demanding on the available bandwidth and it may not be possible to accomplish everything desired to make such plans work. Furthermore, all space-based satellite systems, as well as being costly, remain vulnerable to attack from anti-satellite systems. However, that has not stopped the USA from cultivating cooperation with the UK, Denmark, Greenland and Alaska, Poland, the Czech Republic, Hungary, Romania, Bulgaria, Australia, Russia and Japan. The promise of research and development (R&D) contracts have provided powerful incentives and only Canada has pulled away from involvement.

Stefan Fritsch develops the consequences of this idea further by identifying 'Information and Communication Technologies as a Source of Turbulence'. He uses three theoretical approaches based on IR/IPE (International Relations/International Political Economy) theory, namely: (i) realism/neorealism; (ii) interdependent globalism; and (iii) constructivism, to model the power of modern Information and Communication Technologies (ICTs) to narrow the sovereign political action of most states. Consequently, many nations have lost powers and shed responsibilities to a range of new actors, including multinational companies, nongovernmental organizations (NGOs) and so on. In many senses, the technology is transforming social reality whilst still being dependent on wider social contexts. Fritsch ends with a series of questions about whether or not such techno-processes are deterministic. These kinds of considerations taxed many of the authors in their examination of what pragmatic measures can be taken to manage some of the more negative consequences of modern security becoming ever dependent on bandwidth.

Bruce D. Larkin further explores the consequences of being reliant on an extremely high-tech Global Command and Control System (GCCS) for the launch of nuclear weapons. Although we are asked to believe that such systems are highly protected, impenetrable and foolproof, Larkin points out that other sophisticated US surveillance and communications systems have failed when tested in battle. He cites US aircraft attacking US units working with the Kurds during the last Iraq war. In exploring the prerequisites of the ideal security arrangements for a GCCS, Larkin concludes that cost will be one of the biggest inhibitors in actually meeting the technological require-ments to make any such system failsafe.

Geoffrey Darnton explores the international legal restraints on ambitions to dominate the global information spaces. Darnton finds that although IW is covered by the laws of war in parts, this treatment is comparatively under-developed. How can existing international law be applied to situations and practices that were not even envisioned at the time when treaties, conventions and protocol agreements were being drawn up? Darnton identifies specific treaties that lend themselves to reinterpretation in the light of advances in the role and function of ICTs, but he also identifies many of these as second-order consequences. For example, the development of longer, thinner supply chains and communication lines are more vulnerable than the less efficient but more robust systems that they replace. The breaching of these chains of supply would create serious disruption and potential civil disorder. Darnton concludes that clear improvements are required to provide a proper international framework covering IW but asks, 'Who would enforce it?'

Part III examines differing country perspectives. Whilst much of this book is concerned with the increasingly imperial US perspective of dominating the 'information battle space', it is not surprising that other members of the UN Security Council do not share this supremacist approach and are evolving their own plans along similar lines.

Fanourios Pantelogiannis further explores the situation from the point of view of the Russian Federation. Whilst Soviet thinkers were the first to postulate and analyse the implications of the ongoing RMA, Russia is now a power in decline and their defeat in Afghanistan led to a reappraisal of the country's military priorities. In this case firepower alone proved to be insufficient, and it became more important to invest in command, communication computer infrastructure, surveillance, reconnaissance and electronic warfare. With a sense of being left behind, the former Soviet military commanders looked at new weapons technologies and compared these advances with Russian military hardware, which was rapidly becoming outdated.

Russia is under no illusion that refinancing their information infrastructure, satellites, and so on, is vital if they are not to be left behind. Yet much of Russia's current capacity is old fashioned and decreasing – which impacts on every area of strategic intelligence, including real time communications, early warning and so on. Some of Russia's drive towards refinancing their military telecommunications infrastructure has come from missile and nuclear technology exports although this proliferation to other vying powers (such as China, Iran, Indonesia and India) has not met well with the US who sees it as destabilising.

Russia has also recognised the growing importance of IW as a means to not only enhance the political and psychological impact of its operations but also as a way of increasing the effectiveness and precision of all its available weapons systems. Not all Russian military commanders share these new views. Whether the Russian RMA can be sustained will in part depend upon the availability of adequate funding, as well as the extent to which institutional resistance to reform inside the services can be effectively maintained.

Chris Wu considers the historical development and current priorities of China as it prepares for IW. He describes the new methodologies and systems being developed by the Chinese for waging IW – new advances in radar, satellites and computer systems are being augmented by the development of hard weapons such as killer satellites, electric guns and cruise missiles. Wu also points out the difficulties and disadvantages that China faces in the race to keep up with IW techniques. Lack of training, experience, resources and facilities has meant that China has lagged behind the USA. Indeed, China has, in the past, relied heavily on the results of US experience and products to develop their own systems. Wu notes that this reliance has contributed to the vulnerability of Chinese systems which have little in-built protection against hostile attack. Finally, he outlines a possible scenario of IW tactics that could be deployed by Beijing against Taiwan in which he illustrates a severe disparity in the balance of forces between the two. He poses the question 'How can this disparity be reduced in order to safeguard the security of Taiwan?' and makes two suggestions involving the strengthening of the Taiwanese systems and more direct collaboration with US systems (including anti-satellite systems) to enable them to participate fully in IW at all levels.

The final section of this book deals with the practical challenge of 'what is being or must be done' to manage our individual and collective vulnerability to the threats and opportunities of modern information driven state security systems?

Mike Moore also addresses this autonomous, unaccountable drive towards high-tech dominance as 'A Bridge too Far?' He revisits some of the precision 'bullseye' warfare themes introduced by other contributors and the move by the USA to be the only imperial space power. The roots of this thinking were already evident in the 1950s – but not the satellite technology. Now US military superiority is increasingly turning towards a dogged unilateralism and bilateral arm twisting. R&D contract seduction, identified by other authors in relation to missile defence, is being paralleled by policy laundering in the name of the 'war against terror'. Moore poses the question: 'If one state becomes so powerful globally, how do other states retain full national sovereignty?' This is essentially the running theme of this book. In a time of terror, security debates become increasingly polarized into puerile debates – are you for or against us? Such simplifications engender a climate where technology can be presented as offering something concrete to protect us from a growing international turbulence and sense of decreasing security. If this is a 'bridge too far' then the fact that it is happening against an Orwellian backcloth of permanent war should set alarm bells ringing for the fate of every hard-won limit on the excesses of war and state power which protect us from barbarism. What then should happen next?

Massimo Mauro of the European Commission (EC) looks at the informal cooperation taking place between the European member states and ten Asian countries to further clarify and assess future threats arising from cyber terrorism. In 2002 cyber security was seen as a key priority by ASEM (Asia–Europe Meeting) – the process designed to collectively address these issues, including measures to protect critical information infrastructure and to maintain the balance between the needs of national security and law enforcement with those of the business community which depend on privacy and confidentiality. So far, cyber attacks seem to originate from: so-called 'script kiddies', the lowest form of hacker doing mischief with programmes; financial criminals who penetrate economic systems by stealth, hoping for financial benefits (many of whom are insiders); and political opponents who attack a specific country or organization's website to deny access or use. Nevertheless, Mauro argues that the international community needs to be prepared to meet more professional threats to the Internet infrastructure, individual networks, servers and so on. But the question remains as to who the principal enemy is and what mechanisms are being evolved to rank other national spending priorities accordingly?

Gus Hosein identifies a growing trend for such priorities to be heavily influenced by agenda-setting procedures external to most states. What results is 'Policy Laundering and other Policy Dynamics'. By this Hosein means that policy makers use other jurisdictions to further to their goals. The tactics also include modelling – whereby governments shape laws based on laws in other jurisdictions – and forum shifting – where actors pursue inter-governmental organizations (IGO) rules that suit their interests, then shift to other IGOs when challenges or opposition arises. Hosein sees the emergence of new policy dynamics when national consultative processes either disappear or are severely weakened, with important policy decisions taking place outside traditional democratic institutions. In such contexts, policies are shaped by foreign interests and processes. He provides substantive sections on the CoE convention on cyber crime and the G8 negotiations on high-tech crime. He warns that inter-governmental activities must be paid greater attention since those groups with sway and influence, including external state representatives, are sitting at negotiating tables with unprecedented national powers with the facility to make decisions which are shielded from any parliamentary or democratic process.

The Conclusion by the editors addresses some of the issues for civil society of planned military procurement of information-targeting weapons and systems in the immediate future. The capacity for global telecommunications interception, achieved by the US-dominated Echelon network of worldwide listening posts capable of listening in to all phone, fax and e-mail correspondence, has already caused widespread fears for the future of democracy as we understand it. Already, the national guarantees embodied with various EC member states, in regard to privacy for example, are transcended by Echelon's ability to absorb all entries on the telecommunications highway without as much as a warrant or a 'by your leave'. The European Parliament and Commission has already recognised that such a facility has tremendous implications for the fairness of international economic negotiations, not to mention the manipulation of political discourse which might pre-date any future war and how it is fought.

Because the 'war against terror' is being fought on the basis of intelligence by individuals which can not be checked, the line between information and intelligence is blurred and we begin to experience the start of unaccountable policing. Statewatch in London has been pre-eminent in addressing the extent to which this is already happening. We have seen how IW is still based on the quality of 'information extraction'. Events in Abu Ghraib, Guantanamo and elsewhere have revealed a new willingness to trawl a wider mass of people for potential associates and to torture those incarcerated in the hope of generating further information sources. Some of the commercial telecommunications monitoring systems already on the market such as 'Watson and Holmes' (a telecommunications monitoring system) automatically generate arrest lists from telephone contact chains. It is not hard to see how spurious justifications of association could be generated by telephone tree records to implement extremely repressive actions against specific communities or activists who dissent from the status quo.

Now that both outsourcing of torture and extrajudicial killings are being justified by the war on terror, it is only a stitch away for high-tech digital weapons to become personally targeted onto other digital media held by future suspects, including computers and mobile phones.

The RMA encompasses capabilities yielded by advances in nanotechnology. The book ends with some consideration of how such technology could potentially master us unless adequate checks and balances are put in place to avoid any prejudiced future targeting decisions being made by advanced weaponry on autonomous, algorithmic, self-deciding modes.

Acknowledgements to a Leeds Met student, Professor David Webb and Dr Philippa Trevorrow for the translation of Chris Wu's chapter.

Notes

1. J. Arquilla and D. Ronfeldt, 'Cyberwar is Coming!', *Comparative Strategy*, 12(2) (Spring 1993), 141–65.
2. Colonel R. Szafranski, USAF, *A Theory of Information Warfare: Preparing for 2020* (15 July 2005). Available at http://www.iwar.org/iwar/resources/airchronicles/szfran.htm.
3. For example, C.H. Gray, *War and Computers* (New York: Routledge, 2005).
4. See http://www.dtic.mil/doctrine/jel/service_pubs/afd2_5.pdf (accessed 15 July 2005).

2
Virtual Violence and Real War
Playing War in Computer Games: The Battle with Reality

Martin Bayer

2.1 Introduction

'Sometimes they will give a war – and everybody will join in the game.' Adapting the famous quote of Carl Sandburg might be suitable, according to both the increasing media coverage on computer (war) games and the actual number of games. This chapter will examine how war today has become a sort of interactive entertainment using personal computers and game consoles. It will also address the increasing convergence of commercial off-the-shelf (COTS) war games and military simulations. Further emphasis will be placed on the difference between the alleged 'realism' of such games and the realities of war. Because of the limited scope of this chapter, it will not address the manifold motivations of people to play such games, nor their possible – and widely discussed – effects on players' minds.

For many years, computer games were not particularly well regarded by both public opinion and the established media. Instead of recognising the growing importance of this emerging form of media and even art, computer 'gamers' were either regarded as 'kiddies' or as members of a strange – if not freakish – (male) subculture who never left their darkened rooms and who lived on a diet of pizza and coke. Reality, however, has developed otherwise. Today the computer games industry has a bigger annual turnover than does the movie industry at the box offices. The 'gamer' community is becoming increasingly diverse, opening up to age groups far beyond retirement age, and attracting players of both sexes. The introduction of new games on the market is often accompanied by multi-million pound advertisement campaigns, and virtual characters, such as the ubiquitous Lara Croft, have become twenty-first-century icons.

Computer games, however, do not only feature Italian plumbers with a fondness for jumping over obstacles, abstract blocks to be sorted appropriately, alien invasions to be fended off, or the aforementioned Lara Croft combating spiders and thieves. To an increasing extent, computer games are associated with violence, although only a small proportion of the market has a level of violence that can rival that described in children's fairy tales. However, violence in computer games is depicted not only in fantasy-like environments, but also in 'realistic' settings of war. Due to the rapid rate of technological development, today's computer games offer a breath-taking audio-visual quality, soon equalling contemporary movies. As in the case of movies, this audio-visual quality is both presented and perceived as realistic and authentic. Of course, such a level of mainly graphical 'realism' often has nothing to do with reality; virtual soldiers can carry tremendous amounts of kit without getting tired, virtual wounds can be healed easily, and virtual death is followed by the opportunity to play again. Nevertheless, both movies and computer games will shape the future perception of historical and contemporary wars, especially as the distinction between reality and virtuality becomes increasingly blurred. Additionally, the convergence of commercial computer games and professional military simulations will increase, as the former offers low-cost alternatives to the military, while production costs can be shared. It might be interesting to see the extent to which the future game designer will be 'embedded' into the military, in a manner similar to that observed among news media journalists in recent conflicts.

2.2   Gaming platforms

Computer games are played on a variety of 'platforms'. The first breakthrough into the mass market was achieved with 'home computers' during the early 1980s. Relatively cheap computers like the Commodore 64, the Commodore Amiga, or the Sinclair Spectrum enabled ordinary people (not only) to play games at home. The graphics were fairly simple, as was the game design, but many of these early games were quite addictive. Early war games such as *Blue Max* needed high levels of imagination on behalf of the players, as the representations of 'tanks', 'guns' and 'planes' were restricted to only a couple of coarse pixels.

The core purpose of game consoles, as the name suggests, is playing games. This specialization enables easy handling (for example, there is no need to learn a computer's operation system) and a powerful gaming experience for a modest price (in comparison to a standard PC, especially in the past) or additional mobility (handheld consoles, for example the Game Boy Advance). In some countries, for instance Japan or the United Kingdom, consoles such as the Sony Playstation, the PS2, or the Microsoft X-Box are very successful, having a much larger market share than PC-based games; in other countries, the market shares are reversed. For a long time, most of the console games offered were simple and action-based. With today's consoles, however, more demanding games can be realized. *Medal of Honour: Frontline* is a recent and highly successful example of a next-generation console-based war game set during the Second World War. Game consoles either provide their own small display, or are connected to a TV set. Leaving the future option of HDTV (High Definition Television) aside, the graphical display of console games is strictly limited by the – comparing with today's PC standards – meagre resolution of TV sets.

Increasingly, the PC is the platform of choice for many gamers, especially since today's standard office PCs provide astonishing capabilities for gaming. Specialized input devices such as joysticks, steering wheels or pedals can be attached to increase the realism of the gaming experience. The range of games offered on the PC platform is also the widest: from adventures to role-playing games and from shooters to strategy games and simulations. Since the mid-1990s, PCs have undergone dramatic changes. Approximately every 18 months, the power of the central processing unit (CPU) is doubled. Between 1999 and 2002, the graphics power alone increased tenfold. The development from the first 3D-egoshooter *Castle Wolfenstein 3D* of 1992 (published shortly before *Doom* took the fame) to its 'successor' *Return to Castle Wolfenstein* barely ten years later is amazing. In the former game, a high degree of imagination was needed to regard the blocky blobs on the screen as German soldiers or as attacking shepherd dogs, whereas the latter offers high-resolution 3D-environments with accurate game physics and stunning visuals. One has to bear in mind, however, that in 1992 the visuals of *Castle Wolfenstein 3D* and other contemporary games were regarded as 'absolutely stunning'. This development will continue until the visualization matches the real world – something which will be achieved in the foreseeable future.

The oldest commercial platform for computer games are the 'arcades', expensive devices set up in amusement halls or pubs, where the player pays a small charge for every game – usually lasting a couple of minutes. The grand days of the arcades are actually over, as low-cost platforms like consoles or PCs offer not only a better cost-benefit ratio for people playing games more frequently, but also a wider range of games. Most arcade games are simple action games. The war game aspect is, however, particularly interesting as arcades may offer not only buttons and joysticks as input devices, but also mock-ups of real weapons, such as a sniper rifle in the 'sniper simulation' *Fatal Judgement* ('the game that has all others in its sights', to quote from the manufacturer's advertisement). The Korean game manufacturer Gamebox is offering even more accurate replicas of historical or modern infantry weapons (called *Dreamgun*), for example, the German Second World War machine gun MG-34 or the British SA-80 assault rifle, all with 'shock effect' to mimic the weapons' recoil.

As the visual quality of modern high-end platforms approaches 'real life' quality, there is enough computing power to enrich today's games with physics and 'artificial intelligence' (AI). The former describes the correct representation of physical events, for example a car being driven at high speed into a curve going into a skid. The latter, although being a slight misnomer, encompasses the more or less 'intelligent' reaction of virtual objects to their environment – for example, an enemy soldier who does not wait to be killed, but ducks down after being shot at or after 'seeing' his opponent. To a certain extent, and one might add unfortunately, games are sold and bought with regards to their audio-visual quality, not unlike action movies; impressive explosions, more details, 'realistic' weaponry, and fancy special effects do have some attraction. However, the main issue of a game is entertainment, and it is painful to see how many of today's games fail in this primary objective, while some games of the mid-1980s, despite their technological limitations, still provide rewarding gaming experiences.

One brief outlook of the future might resemble some descriptions out of Ray Bradbury's science fiction novel *Fahrenheit 451*, in which 3D 'TV-sets' occupied several walls of the living rooms; in computer-aided virtual environments (CAVEs), 3D images are projected on as many as six walls. Standing in the middle of this room, the gamer does not need to wear a clumsy VR (virtual reality) helmet, but only a pair of lightweight polarizing glasses, a first step towards the *Star Trek* dream of a 'holodeck'. Thus, the player can move through the virtual environment without the need of a screen, a keyboard, or a mouse, making the whole experience a much more realistic one. Such CAVEs do already exist, the first being developed in 1991. It may be only a few years before the first game CAVEs are cheap enough to provide for a viable business model for amusement halls. However, remembering that some people already have problems distinguishing the virtual from the real world (although this phenomenon is still mostly connected to other media, such as TV and movies), such a technological development may lead to serious problems with reality for some unstable minds.

2.3 Definition and historical context

In order to define computer war games it might be easiest to start by explaining what they are not. Computer war games do not deal with fantasy environments ('dungeons and dragons') where the use of weaponry is enriched by magic spells and other supernatural powers, and where players battle against hordes of zombies, orcs, dragons, and similar creatures. Additionally, science fiction environments are also excluded from this definition, however 'accurately' the fighting in space against alien races may be simulated. Thus, computer war games are about simulating or re-enacting historical or contemporary conflicts, including hypothetical events to some extent. Nevertheless, there is a grey area in this definition; while the hypothetical events displayed in *Operation Flashpoint: Cold War Crisis* might be regarded as connected to a possible reality, the zombies revived by mad Nazi scientists and leather-clad SS dominatrices in *Return to Castle Wolfenstein* are clearly not. Another example for 'what if?' cases, is the ego-shooter *Iron Storm*: in this game, the First World War has not ended in 1918, but has continued for another 50 years. It is very important, after all, that such games are entertainment, to which sometimes a (usually political) message has been added. Another aspect is the ongoing convergence of commercial off-the-shelf (COTS) computer games with professional military simulations.

Computer war games are set in a variety of historical contexts. The Stone Age and the era of Classical Antiquity are represented by games such as *Age of Empires*, where the player can even develop his 'race' from the former to the latter, that is to the Iron Age. In games set in the Middle Ages, such as *Stronghold* or *The Age of Kings*, building and battling castles is usually the focus of the game. In these games, a strong emphasis is set on siege engines, which is not historically accurate; in reality, sieges were usually based around starving out the opponent – an option which would be boring and not very entertaining if put in a game. The early modern period is addressed by games such as *Cossacks* or *Shogun*. As soon as firearms are introduced as weaponry, the guns' ranges often differ from their real-life counterparts in order to balance the different units, thus producing a more challenging game (and a less accurate depiction of historical events). Inevitably, the modern period is most interesting for game developers, offering different weapon systems and all kinds of theatres of war. The First World War is relatively seldom represented, featuring mainly in flight simulations such as *Dawn of Aces*. This is perhaps unsurprising as the characteristic methods of warfare in this conflict were shelling the trenches, storming against barbed wire entanglements, and being machine-gunned down; it is hard to imagine a thrilling and entertaining gaming experience based on historical accounts.

The 'mother of all wars' for computer war games is surely the Second World War; there is everything a player could want to have for such kind of entertainment. Firstly, there is a clear distinction between good (heroes) and evil (villains), although choosing the 'evil' may offer a special thrill for some. Secondly, there are huge varieties of weapon systems, including planes, tanks, submarines, surface vessels, and all sorts of infantry weapons. Fighting can be done on the ground, in the air, and on and below the water surface. Thirdly, the theatres of war portrayed in these games range from the desert of North Africa to the Atlantic, from the Pacific Islands to South East Asia, from Norway to Italy, and from the UK to the vast steppes of the Soviet Union. Thus, players who prefer 'exotic' environments are served, as are those preferring 'local' environments, such as their own country or city. The list of games connected to the Second World War is almost endless. Recent examples include the successful *Medal of Honour* and *Battlefield 1942* series (ego-shooters), the flight simulator *Il-2 Shturmovik*, and the strategy games *Blitzkrieg*, *Hidden & Dangerous II*, and *Combat Mission*.

The Cold War is seldom the setting for computer games, although, recently, quite a few games – such as *Vietcong* – have been based on the Vietnam War. Most proxy wars seem to lack the excitement needed for a game, while the main actors of the Cold War did not fight each other directly (at least not on a large scale). Thus, games using the Cold War as a background usually do so on the basis of hypothetical events. The story of the highly realistic ego-shooter *Operation Flashpoint: Cold War Crisis* is based on a Soviet 'rogue' general invading the fictitious East-European country Everon during the mid-1980s after Gorbachev had become president.[1] Games covering the contemporary world often use a similar approach; for example, *Conflict Zone* or *Real War* offer fictitious enemies like terrorist organizations, international companies or state alliances, challenging the North Atlantic Treaty Organization (NATO), the United Nations (UN) or the West in general. *World War III – Black Gold*, in which the player could choose from playing either the US, Chinese, or Iraqi side to fight for the world's last oil reserves, seems to be rather outdated now, or highly relevant, depending on your point of view.

As in today's real world, terrorism is now a hot topic for games. Shortly after September 11, a few simple, homemade games appeared on the Internet, one such being *Bin Laden Liquors*, in which the player could kill a virtual Osama bin Laden. In an increasing number of games, such as *Splinter Cell*, the *Rogue Spear* series, *Soldier of Fortune II*, and *Land Warrior III*, the fight against terrorism can be pursued on the home front. While the viewpoint put forward in these games was predominantly a western one, an increasing number of games are being produced by sections of organizations such as Hamas or Hizbollah; games such as *Underash* or *Special Force* enable the player to fight against Israeli soldiers. Like *America's Army*, those games ultimately try to recruit new soldiers for the 'real thing', thus opening a new dimension to gaming, the political background. Arguably, games such as *Back to Baghdad* (1994) always had a political aspect,[2] but it needed the 'recruitment games' to implement political ideas on a much larger scale. It will remain interesting to see how the political content of (especially war) games will develop over the next few years, particularly in the non-western world.

2.4 Computer game genres

Nearly all genres of computer games (except sports and racing) can feature warfare as their main subject. In adventures, the player is using a pre-defined character solving a set of puzzles, eventually accomplishing the main objective, often by interacting with other people (that is, virtual characters). The main objective might be as limited as rescuing a princess or as broad as saving the world. Usually, adventures are single-player games only. One of the many war games of this genre is the recent 3D-action adventure *Prisoner of War* (PoW), where the player, as a US Air Force officer held captive in various PoW camps in Nazi Germany, has to solve puzzles to free himself. However, as this would obviously not be enough to make a 'higher goal', he also has to find information about Germany's programme to develop 'weapons of mass destruction'. Through accomplishing this modern-sounding mission, he ultimately saves the world.

Role-playing games (RPGs) are to some extent similar to adventures. The player also has to solve a set of puzzles; however, in RPGs the fighting element is much more important. Additionally, the player does not only control one pre-defined character, but a whole set of characters whose abilities (for example, the use of specific weapons or sneaking) can improve throughout the game. Most role-playing games are set in fantasy worlds, but there are some war games – for example, the *Jagged Alliance* series – where the player commands a group of mercenaries. A fascinating 'real-life approach' might become the new RPG by the US Army, *America's Army: Soldiers*, which is soon to be published. In this game, the player can develop a career inside the US Army and can learn about the organization and possible career prospects. This emphasis on 'educational' aspects is not surprising because *Soldiers* is a 'recruitment game' and is thus distributed without any charge.

Simple action games, also called 'shoot' em ups', are now increasingly out of fashion. Regarding war games, the mere content of killing as many waves of enemies in as short a time as possible seems unappealing to many people. Secondly, it regularly led to *Indizierung* ('indication') in Germany, the third-largest PC games market world wide, because of its glorification of war and violence.[3] An example is the old C-64 game *Commando Libya*, where the player's only task was to machine-gun waves of 'Ghaddafi's children' charging from the top to the lower end of the screen. Often, no background story is provided, it is just faceless enemies without any affiliation to a country or a cause that need to be killed in order to score points.

A very important genre within war games is simulation. The mastery of complex technology is prominent; the actual war fighting may even be regarded as less important. Nearly all kinds of equipment have already been simulated (with an emphasis on flight simulations), from tanks (*Armoured Fist*) to submarines (*Aces of the Deep*), and from helicopters (*Gunship!*) to planes (the *Combat Flight Simulator* series). There is, however, a distinction between highly accurate and 'action-oriented' simulations. The former group, consisting of games such as *EF2000* (simulating the Eurofighter *Typhoon*), need considerable practice time in order for the player to 'master' the machinery. By contrast, those games in the latter group place a greater emphasis on the 'fun' element, which is, in the case of combat simulators, the actual fighting. The equipment is not simulated to its full complexity, leading to a shallowes learning curve. An example of such action simulations is the *Comanche* series, which features the RAH-66 helicopter. While some more realistic simulations also include the possibility of malfunction (for example, torpedo duds or missiles missing their designated targets), less complex games may lead to a certain kind of (mis-)perception: in such simulations, precision weapons always hit their target, which is not the case even with the most sophisticated systems available to the military. Gamers may recognise 'their' hardware in media reports on actual conflicts and wrongly apply their 'experience' to reality, for example the near-invincibility in some games. Furthermore, the widespread suffering caused by war is not taken into any account, which often results in a 'clean' and chivalrous form of fighting 'man vs man', especially in historic flight simulators. An important aspect is the convergence of professional simulators with COTS simulation games. Microsoft's *Flight Simulator*, for example, is being used by the US Armed Forces for theoretical flight training, since it combines an adequate level of realism for the task with cheap licensing fees, especially in comparison to 'professional' simulations.

One of the most popular genres in contemporary gaming is the so-called 'strategy game'. There are many non-violent strategy games, in which, for example, the player has to develop a city's infrastructure (as in the *SimCity* series). Nevertheless, many strategy games focus on military content. To dispel a common misunderstanding at the very beginning, despite their name most strategy war games are tackling not the strategic but rather the tactical level of warfare. Single tanks or soldiers (or at maximum a group of them) are commanded, which leads in some cases to a frantic micro-management. There are two main strands – the first (and older) consists of 'turn-based', the second (and more popular) of 'real-time' strategy games (RTS). The former may be described as the continuation of chess by other means, being based on board and pen-and-paper games. Two (or more) armies fight each other, whereas each player may carry out a certain number of moves during each turn, for example moving a tank to a certain position, and, if there are enough 'action points' left, firing, repairing, or rearming it. Some of these games go into a very deep level of simulation, for example *Eastern Front*: all infantry weapons, tank version, or truck types of the historical conflict at a certain time and place are included, as are many hampering factors, such as the terrain type, its height, the overall visibility, or the weather. Less accurate but more playable are games such as the *Panzer General* series, whose first instalment was put on the index in Germany, as the player had to wage a war of aggression.

Since the early 1990s, RTS games have become one of the most successful of all genres. The main feature of turn-based strategy games is that all of the players' actions (including the virtual ones) are undertaken without having alternating turns, but at the same time. This leads to a faster-paced game, which seems to appeal to many gamers. However encompassing the genre description may be, the range of approaches adopted by the individual games is quite noteworthy. In *Commandos II*, for example, the player must not lose a single one of his special forces soldiers while undertaking missions behind German lines during the Second World War. In contrast to most RTS games, the player cannot 'produce' more units and eventually crush the opponent by numerical superiority (the so-called 'tank rush').

More typical examples of such RTS games are *Conflict Zone* and *Real War*, in which the player can produce his units (also including soldiers) by building factories or barracks, and using resources such as money, oil, or his standing in public opinion. In this respect, the *Sudden Strike* series uses a more realistic approach, as the player may receive some reinforcements later in the game, but he cannot 'build' new soldiers or equipment. Nearly all of these games use the so-called isometric perspective (a view from above at an angle) usually accompanied by several zoom levels. The number of units represented may range from only a handful, as in *Commandos*, to several hundred, as in *Blitzkrieg*, leading to different kinds of demands on the player.

A very prominent genre, but by no means the dominant one, is the ego- or first-person shooter (FPS). In such games, the player sees the virtual 3D world 'through the eyes' of his character, while holding a weapon (seen on the lower part of the screen) and blazing his way through numerous opponents. The player simply has to move his character over weapons and ammunition to supply him with the hardware needed, which are scattered throughout the maze's rooms. If the hero becomes wounded, 'health packs' can be picked up for some instant relief. *Castle Wolfenstein 3D*, launched in 1992, was the first of these games, but one year later the genre's breakthrough came with *Doom*, as normal office-PCs became powerful enough to be used for 'real' gaming (that is graphics-based games). Although the first representative of ego-shooters was a war game, it took some years to see another example, while science fiction or fantasy-based environments dominated the genre. In recent years, however, war games have triumphed over ego-shooters. Until the mid-1990s, both the environment and the level objectives were quite simple: find the exit and kill everything that is moving. Since this time, the genre has developed tremendously, allowing scope for complex scenarios and cooperation. As with simulations, there are, on the one hand, more realistic ego-shooters such as *America's Army: Operations* or *Operation Flashpoint*, and, on the other, action-oriented games such as *Medal of Honour* or *Battlefield 1942*. The former put emphasis on the 'realistic' experience of fighting, requiring the player, to hide, wait, save ammunition, and proceed in a military tactical way. The second group of shooters place the emphasis on the action: there is lots of available ammunition, the opponents to be killed are numerous, the player can carry many weapons and may be healed instantly, while tactics like jumping and circle-strafing (moving sideways around an enemy while shooting) make sense, in contrast to the real world. Many of today's ego-shooters are multiplayer games, where two parties of four, eight, or sometimes more players fight against one another. A well-known example is *Half Life: Counterstrike*, a multi-player game developed by gamers, based on the commercial *Half Life*. In this game, a group of terrorists are fighting against a group of policemen. To be successful in this virtual version of 'cops and robbers', the players need to communicate extensively and to coordinate their tactics effectively.

2.5   Realism versus reality

As with the movie industry, the desire for 'absolute' realism in gaming is high; often, it is used in advertisement campaigns as a key selling point. However, what kind of realism is desired? To approach this question, a short case study might be useful: 'You don't play – you volunteer'™ is the marketing slogan for the Second World War shooter *Medal of Honour*. Interestingly, this project was initiated by Steven Spielberg and his Dreamworks Studios after their production of the war movie *Saving Private Ryan*, a film that set new levels of (perceived) realism and tried its best to involve the audience in the action, using audio-visual effects, newsreel-style camerawork, and the realistic depiction of mass violence. In *Medal of Honour*, Spielberg wanted to create an interactive experience that would be as close to the movie as possible. The developers did their very best (including the assistance of the same military advisor who once worked on *Saving Private Ryan*), and both *Medal of Honour: Allied Assault* and *Medal of Honour: Frontline* provided an astonishing level of audio-visual quality. Both games are highly successful and very playable. As with the movie, the game's most stunning sequence is the invasion of the French Normandy coast on D-Day, 6 June 1944. In this first level, *Medal of Honour* becomes realistic in both a fascinating and frustrating way, since the player's only task is to survive. He cannot defend himself actively, but only by ducking down and running to the next covering. While doing so, the player passes scenes of wounded soldiers, of panicked soldiers refusing to move, and of bodies blown up, and it is very likely that he will be killed by one of the numerous bullets coming in from the far distance. It is the interactive re-enactment of a re-enactment (*Saving Private Ryan*) of the Omaha Beach landing. Usually, game developers do not willingly expose their audience to such high levels of frustration, as surviving is just a stroke of luck. This is about where the 'real realism' stops. Most war games, although they claim to be 'realistic', offer quite strange definitions of 'realism'. This should come as no surprise, as games are intended to entertain, to be fun, and the utmost level of realism can only be boring, frustrating, disgusting, or a mixture of these or more negative factors. To quote the producer of *Medal of Honour: Allied Assault*: 'We are putting our focus into authenticity, not necessarily total realism. We want the game to be as enjoyable as possible, and try not to sacrifice fun at the expense of accurate physics or ballistics.'

To distinguish between 'realism' and reality, some aspects of simulated and real war fighting need to be considered. As in action movies, weapons are very prominent in war games. Developers announce on their websites that are using original historical weaponry to make their virtual counterparts 'authentic'. Today, gamers can expect that the different weapons in the game will sound like the real ones, as the sounds of loading, cocking and shooting have been sampled. The distinction between game and reality usually begins with the use of these weapons. Firstly, in many games, the player can carry an impressive amount of kit. For example, in the aforementioned game *Medal of Honour*, he may carry a pistol, a sniper rifle, a machine pistol, an automatic assault rifle (which were quite heavy during the Second World War) and a bazooka, accompanied by over 1,200 rounds of all calibres involved and a couple of hand grenades; in real combat, no soldier would be burdened with this weight, and he would also struggle to pack all of these items. Games such as *Operation Flashpoint* or *Rogue Spear III* are the exception, since in these the player is limited to a primary weapon and an optional secondary sidearm, with only a reasonable amount of ammunition. Of course, this limitation changes the whole character of the game: if a soldier is only equipped with an assault rifle and 60 rounds of ammunition, he may not engage a whole enemy company. In action-oriented shooters, however, firing thousands of rounds(!) in each level is no exception. For example, the magazine *Playstation Planet* wrote in its review of *Medal of Honour: Frontline*: 'Without a doubt, *Medal of Honour* is the best and *most realistic* war game you can buy ... You'll be so busy pulling the trigger, you won't notice what time it is' (emphasis added by the author). So much for 'realism'. But the primary action for soldiers in war is waiting (and trying to survive), which is hardly a thrilling experience for gaming.

The way in which these weapons are used is also significant. In most first-person shooters, the weapon chosen is shown in the lower part of the screen, with a crosshair in the middle of the screen providing the weapon's aiming point. In many games, the player will fire his weapon, resulting in an exact strike on the object in the crosshair, even for several consecutive shots. Similar to Western movies, the weapons are more or less fired from the hip, and there is also no recoil to be taken into account. More realistic are games such as *Operation Flashpoint* or *Vietcong*: firing from the hip may be possible in these games, but hitting something is highly unlikely. In order to take proper aim, the player needs to look through the weapon's sight, thus restricting his view. Even more so, not only is each shot being followed by recoil, but also the player's breathing is taken into account, forcing him to choose the right moment for pulling the trigger. Nevertheless, this is still by no means a full shooting training, as most real-world mistakes are connected to using the sights or pulling the trigger too fast. Only virtual military shooting ranges, such as the AGSHP of the German Army, using modified infantry weapons, provide for such a realistic training. It is noteworthy, however, that games companies such as Bohemia Interactive (the Czech producer of *Operation Flashpoint*) are also offering military simulators. Surely, the convergence of games with military applications will increase.

Speaking of the soldier, human movements and behaviour are far from being adequately simulated in these games. Usually, there is no fatigue, and the virtual soldier may run endlessly without getting tired. Often, movements are very restricted for the sake of an easy interface and to ensure high levels of playability. Nevertheless, it is somewhat odd when a soldier cannot just look around the corner without exposing his full body, which is still the case in many shooters. When moving around in the game, many 'natural' ways are blocked – for example, a soldier may not cross through a hedge or jump over a small ditch, simply because he is not allowed to do so. Additionally, in games, war is always 'action', but in reality, most of the time the soldier spends is waiting or doing repetitive, if not boring tasks like cleaning his weapons, or guard duty. This is especially relevant in respect of 'recruiting games' like *America's Army*, since it may lead to a misperception of life in the military on behalf of the gamer, which will be rapidly cured if he should eventually enrol for military service. An interesting side issue may be called 'racial equity', as *America's Army* is the first game in which the player can choose from different skin colours to fit Asian, Afro-American, Latino, or Caucasian background.

Even stranger than the restrictions on the player is the behaviour of many enemies: they just seem to long to be killed, as taking cover or firing whilst unexposed are obviously impossible skills for many virtual enemies. However, one has to add that over the last few years, the 'artificial stupidity' of enemy behaviour has been reduced to some extent, but 'artificial intelligence' is still far away from being a literal use of the term, thus being responsible for one of all these games' main weaknesses.

One prominent aspect of wargaming is the depiction of wounds and death. There are four different ways of depicting wounds in games. The first one is not to show any, being used, for example, in *Medal of Honour*. Soldiers may walk slower or limp, but there are no pools of blood, no severed limbs, and no nasty wounds. Interestingly, this approach seems to be the one favoured by most gamers, as they play the game not to wound, kill, or have fun watching others suffering, but to succeed in the missions by communicating and acting in concert with their team members. Most *Counterstrike* players thus switch off the option to depict blood and gore. The second strand, taking *America's Army: Operations* as an example, depict a paintball-like red blob on the uniform if a soldier has been hit. The third strand depicts violence and wounds in a more or less 'realistic' way. In *Operation Flashpoint*, the uniform gets virtually soaked with blood where the hit(s) occurred. The fourth way is fortunately used only very seldom: games such as *Soldier of Fortune II: Double Helix* claim to only provide 'realistic damage effects', but in fact they offer a gory experience, arguably with the intent of attracting users with this 'open' approach to violence. As a result, it has not only been criticized in the game reviews, but also by the audience. Such games are highly unpopular with most gamers, as their focus is on the gameplay, and not on shooting each other's limbs off, not to mention hundreds of frames of animation if somebody took a shot in the balls. Generally, 'clean' wounds from gunshots prevail in games; nobody gets maimed, turned inside-out, or loses one's limbs. On battlefields, however, artillery shells and bombs are usually causing the highest rate of casualties, including most of the nasty wounds. Nevertheless, similar to most war movies, people die quickly, if not heroically.

Not only the depiction of wounds, but also their effects and their treatment are interesting aspects. In most games, the player can instantly improve his health or even completely heal himself by using 'health packs' which can be found throughout the game. In action-oriented shooters, it is quite normal to be hit by 150 or more bullets during a single mission; no problem, as long as there are enough health packs; of course, there are also no long-term effects. Games such as *Operation Flashpoint* do not provide such treatments: if the player is hit, he is highly restricted (depending on where he took the bullet). Another 'arcade issue' is that often the higher rank an enemy has, the more hits he can take, which does not really correspond to real life, where a staff officer will usually wear no more body armour than an infantry soldier.

After the 'health level' of a virtual soldier has reached 0 per cent, he dies. Until then, the player is often not restricted in his movements or his speed despite being badly hit. Even with only 1 per cent of his strength, the player may run and jump around. In most games, killed people literally just 'drop dead', a couple of seconds later they vanish, especially in action-oriented games. It would probably be just too ridiculous to walk on these piles of corpses, because in a single level (lasting maybe a quarter or half an hour) killing 100 or more enemies is quite common. Of course, the ultimate difference from reality (and the most obvious one) is that there is always the possibility to restart a game, which will provide no problems as long as there is a clear distinction between real and virtual world.

Interestingly, taking prisoners (or not wiping out the enemy) is not an objective to most games. Extremely seldom, missions include the possibility of taking an enemy prisoner. In most games, however, enemy soldiers neither give up nor flee. Even if totally outnumbered, they just keep on fighting until the last one is killed. This 'total war' approach of killing everybody may have its parallel to the 'blood rage' occurring in real warfare; the non-existence of surrender and PoWs (except the friendly ones the player may have to liberate) is surely strange. There is no way of negotiating, no other possibility than to kill the opponent, the more the better, and in many games it is even necessary to destroy all enemy units in order to complete a level or to receive a medal.

Civilians do not appear in most war games, similar to many war movies. If they are depicted, they might be mad Nazi scientists (who are to be killed anyway), but not normal people inhabiting the fighting zones. This is even stranger in scenarios of urban warfare, if one remembers the contemporary problems with civilian casualties, or the high number of civilians killed in all conflicts since the start of the twentieth century. Some games try to give an explanation; for example, in one mission of *Medal of Honour: Allied*

Assault, set in a half-destroyed French town, the briefing says that 'all civilians have been evacuated beforehand'. Of course, as there are no civilians, there is also no need to bother about unintentionally killing them or even to consider civil–military relations during military operations.

The so-called 'collateral damage' has been a hot topic for quite a time, and the influence of the media on the outbreak and course of wars (for example, regarding popular support for a campaign) is well known. In war games, however, these issues hardly matter, with the exception, for example, of the game *Conflict Zone*; if the player chooses the 'good' side (an organization similar to the UN), civilians have not only to be spared but protected and eventually evacuated from the war zone, and collateral damage has to be avoided. However, if the player chooses the 'evil' side (a conglomerate of rogue states and international groups of companies), he may use media for propaganda purposes instead.

Logistics operations, an enormous part of any military activity, in terms of both importance and expenditure, are seldom a significant factor in war games. In many strategy games, new units can be easily repaired and even produced. To an increasing extent, real-time strategy games try to add a realistic logistics aspect, for example by not offering such fast production of additional forces, and forcing the games to deal with basic supplies. However, tanks and other weapons often do not run out of ammunition or fuel, and even if this aspect is regarded at all, there may be just a game penalty – for example, the loss of 'action points' during a turn.

Precision regarding both small arms and larger weapon systems is another issue. Only the more proficient simulations include the possibility of malfunctions. In most cases, a locked-on missile will always hit its designated target. In some team-based shooters, for example in the recent *Conflict: Desert Storm*, the hit ratio of the friendly soldiers is exaggerated, and enemy equipment can be destroyed in almost arcade-like fashion. It could plausibly be argued that one is more likely to believe the reports on the successful use of precision-guided munitions in real conflicts if one has 'used' the same equipment with a similar positive outcome.

2.6 Games and professional military simulations

The convergence of COTS games and military professional simulations will continue. The US Army has already developed the aforementioned *America's Army: Operations* as a 'recruitment game'. More than 700,000 people downloaded the free game in the first couple of weeks, while several millions of CD-ROMs were distributed. In just a couple of months, it had become one of the most successful multiplayer games on the Internet, perhaps also because it is free of charge while offering good graphics and gameplay. Some 25 per cent of all requests for information on becoming a soldier in the US Army are now received via the *America's Army* portal website – a huge success, contrary to the belief of many recruiters, who despised the game as humbug until it succeeded so well. Furthermore, two other games are in production for the US Army by the games developer Pandemic Studios. One of them, *Full Spectrum Warrior*, is a tactical military operations in urban terrain (MOUT) fighting game, which will not only be published as a commercial game, but will also be used, in a slightly adapted version, to train infantry soldiers. Other nations follow track. The British Army, for example, has 'militarized' the shooter *Half Life*. As the software interface for add-ons are well documented, and the producers of such games encourage their users to develop their own scenarios, this approach is by far the cheapest one; in this case, the adaptation did cost less than €80,000, a fraction of the costs to produce a new game. While in the media, the image of such games as 'killer trainers' is often perpetuated, the military does not use such games for training its soldiers to fight, become more aggressive, or even effectively kill their enemies. The main purpose is instead to train communication skills within the teams, to foster teamwork, and to improve procedural training – for example, how to clear a building. In reality, however, it is far less likely that operations would be conducted in this way, as the danger of own casualties in MOUT, also due to 'friendly fire', is just too high.

The interest of the military in developing better simulations to train soldiers is high, especially, unsurprisingly, in the USA. Furthermore, computer games have been employed as part of the training environment. There are already some conferences focusing on games for 'serious' purposes, such as the *Serious Games Summit*. While many industries and the medical services are also interested, the main impetus comes from the military. Computer games have various advantages. Firstly, they are easily deployable: briefings and trainings can be carried out anywhere without additional equipment, an important aspect for overseas missions. Secondly, additional missions for the games can be easily developed, ideal for changing environments. Thirdly, games are regarded as being attractive to future recruits. As Dr Michael Macedonia, Chief Scientist at STRICOM[4] of the US Army, put it: 'the recruits want to have some vicarious excitement, and, damn, we can give them some vicarious excitement.'[5] Additionally, computer games can train communication and cooperation skills, situational awareness, and even cultural awareness – these skills are essential for fighting in urban areas, or for peacekeeping and peace enforcement missions in foreign countries in general.

Given the developments above, computer games should have a bright future in the military. One may think of controlling the ubiquitous unmanned aerial vehicles (UAVs), that now come with a 'C' for combat (as UCAVs). Hard- and software to command these drones resemble computer simulations. Right now, some ten soldiers are needed to control a single *Predator* drone; this is to be reduced to one operator per five drones in the near future. To cope with these demands, having experience with computer games is surely advantageous. Of course, there are also negative aspects. While the present

and future generations of recruits may be very aware of information technology and games, on average, they lack physical abilities. The armed forces,
however, are of little use to coach potatoes. Additionally, there is the danger
that successful gaming experiences are wrongly transferred to real military
actions, without noting that reality and virtuality are two different sets.
Enemy artificial intelligence can only give an idea of the possible actions of
human opponents, who tend to be very creative to overcome their limitations, for example in heavy armour and high-tech material.

According to many media reports on violent computer games, there is
another issue which could be exploited by the military: these games, especially realistic war games, are depicted as 'killer trainers' that reduce the
threshold of killing human life by habitualization and training, for example
to exterminate thousands of virtual enemies, rewarding head shots with
additional points. This argument sounds plausible, as the environment of
these games are realistic war or anti-terror operations, and the fact that the
armed forces are using these games seems to support it. However, as with
games, the reality is different. Firstly, the armed forces of the world have
much better and proven means of reducing the threshold to kill *when acting
as a soldier in a war environment*, for example, close-combat training or
bayonet exercises. Secondly, the armed forces are using these games to train
communication, cooperation and situational awareness, together with a few
other (dominantly 'dual-use') skills, and not for training to shoot or even
kill. Moving a crosshair on a computer screen by a mouse and firing a
virtual weapon by a mouse click is completely different to aiming and firing
a real weapon; not only in terms of technique, but also regarding the
threshold to fire it at a target – especially a human one.[6] Thirdly, and arguably most importantly, according to numerous studies on human behaviour, players of all ages can distinguish between virtual and real realities,
with the exception of people with severe personality disorders. This will
continue to be the case, as long as the players interact with the virtual
reality via computer screens and interface devices such as a keyboard and
mouse. Perhaps this chapter will have to be rewritten when a true 'mind-
link' – a direct interaction between the human brain and a computer – have
been designed. To date, however, it is evident that one cannot be trained to
kill simply by playing computer games.

2.7 Conclusion

Overall, most computer war games are completely unrealistic, despite their
stunning audio-visual aspects. The latter will surely continue to improve,
soon providing almost-lifelike quality with standard home PCs or the next
generations of game consoles. Although games such as *Operation Flashpoint, America's Army: Operations,* or *Full Spectrum Warrior* prove that it is
possible to produce a both gripping and (relatively) realistic war game, any

further 'improvements' in realism are probably undesirable, since they might mean that games become overly complex, boring, and frustrating for players seeking entertainment. Furthermore, the great majority of gamers do not wish to see a realistic depiction of violence, wounds, and death. Therefore, any virtual 'experience' of war in computer games is likely to be a very limited one, even if future visual application systems are taken into account. This is by intent and no surprise, since the aim of gaming ultimately is entertainment and fun: 'The entertainment industry creates memories you keep for life, we (the military) want to create memories you should forget about but learn from.'[7] One may ask, however, to what extent an inherently violent activity such as war can be regarded as fun at all. Firstly, one has to remember that the majority of computer games do not consist of violent or even war games – even if reports in the media present a very different picture. Secondly, war always has been – continues to be – a central human activity with its own, powerful, fascination. As within all other forms of media, war will continue to play an important role in computer games.

Media attention on violent computer games is particularly high at present, especially following incidents such as the Columbine High School killings in 1999 or the Erfurt massacre in 2002. Games such as *Counterstrike* are frequently depicted as 'killer trainers', particularly since the military does indeed use adapted versions of COTS war games for their own training purposes. Many people assume that if the military, being in the 'real business' of killing, is using such games, there must have been some influence on the killers, either by reducing the human threshold to kill another of its own species through habitualization, or simply by improving their aiming and shooting skills. In particular, the former US military psychologist Dave Grossman attacks arcade video games which employ mock-up models of weapons. He argues that habitualization is indeed happening, and that players of violent games are more inclined towards violent behaviour – if not even killing human life – in reality. However, one has to take into account that all of these young killers had also had both access to and received training with real guns. Additionally, although several armed forces are using COTS computer games for training purposes, these games are neither intended to reduce the threshold to kill (the military has much better means for doing this) nor to improve skills other than communication, cooperation, and team spirit, all of which are definitely of 'dual use' and positive for civilian life.

Given the continuing high speed of technological advance, one can clearly envisage the development of interactive entertainment in the future. Both the 'holodeck' and the 'mind link' (providing for imagery and emotions directly within the human brain) are still science fiction, but as both entertainment and war are large markets, we may well see their application in our lifetimes. In addition, the possible influence on the perception of historical and contemporary events through modern media (including their wide range of possibilities of both faking and mimicking reality) should not be underrated. For the younger and upcoming generations, movies such as *Saving Private Ryan* and related games such as *Battlefield 1942* and *Medal of Honour* may represent the reality of events; they look just too convincing, and the alleged high level of realism is being advertised as a proof of their quality. The already high responsibility of media producers, thus, may increase still further in the future. Nevertheless, one should not expect computer games to provide their users with thorough explanations for the causes of wars, the historical and sociological backgrounds, and the motivations of their protagonists. This lack of factual information and the presentation of both easy (military) solutions and clear-cut distinctions between good and evil may be less due to the usual tendency of games to simplify, but more a derivation of the political reality. The advertisement campaign for the Second World War-strategy game *Blitzkrieg* puts it quite simply: 'Play war, don't make war.' Even if real war as such will not be regarded as a 'normal' kind of activity (beyond the levels of contemporary politics) or even fun, truth will definitely become an increasingly precious value. Distinguishing reality from virtuality may become an essential feature in the near future.

Notes

1. It is interesting to note that the game producer is also producing simulators for the military.
2. According to an advertisement for the game, 'You are going Back To Baghdad™ to finish the war that George Bush prematurely stopped. You're the flight leader of an F-16C Block 50 [a multi-role fighter aircraft] with all the armament needed to get the job done.'
3. Germany has quite strict laws regarding depicting violence in media. Computer games need to be presented for approval by an official committee, otherwise they are regarded as a product only for adults, even if the content is intended for children and is comparatively harmless. If games such as the *Quake* series are perceived as focusing on violence as their main if not their sole content, they are placed on an index. Such games must not be advertised, and they may only be sold to adults 'under the counter'. While in the first years of computer gaming, this index was seen as a reverse 'seal of approval' by many gamers, the ban on advertising a game damages its market position in today's world of huge production costs and high levels of competition. If a game depicts Nazi symbols or glorifies Nazi ideals, it is completely banned and cannot be distributed at all.
4. STRICOM: US Army Simulation, Training and Instrumentation Command, based in Orlando, FL.
5. Conference presentation, Defence Simulation and Training, London, 7 November 2002.

6. According to a study conducted by the US Army, conducted during the Second World War, only about 25 per cent of US soldiers fired *directly* at the enemy, even while under attack, whereas the majority either did not shoot at all, or only fired in the direction of the enemy. As a result, basic training was reformed to reduce the high threshold to kill.
7. Dr M Macedonia, conference presentation, Defence Simulation and Training, London, 7 November 2002.

Bibliography

Anderson, C.A. and K.E. Dill, 'Video Games and Aggressive Thoughts, Feelings, and Behaviour in the Laboratory and in Life', *Journal of Personality and Social Psychology*, 78(4) (April 2000), 772–90. Available at http://www.apa.org/journals/psp/psp784772.html.

Albrecht, H. 'Blut und Spiele', *Die Zeit*, 19/2002. Available at http://www.zeit.de/2002/19/Politik/print_200219_computerspiele.html.

Büttner, C. 'Zum Verhältnis von phantasierter zu realer Gewalt' (1996). Available at http://www.bpb.de/snp/referate/buettner.htm.

Demaria, R. and J.I. Wilson, *High Score! The Illustrated History of Electronic Games* (Berkeley, CA: McGraw-Hill/Osborne, 2002).

Der Derian, J. *Virtuous War: Mapping the Military–Industrial–Media–Entertainment Network* (Boulder, CO: Westview Press, 2001).

Eng, P. 'A Play for Better Soldiers – The Rise of Computer Games to Recruit and Train US Soldiers', *abcNews.com*, 21 August 2002. Available at http://abcnews.go.com/sections/scitech/DailyNews/wargames020821.html.

Fritz, J. and W. Fehr, 'Computerspiele zwischen Faszination und Gewalt' (1996). Available at http://www.bpb.de/snp/referate/fritzst8.htm.

Fritz, J. and W. Fehr, 'Gewalt, Aggression und Krieg – Bestimmende Spielthematiken in Computerspielen', in J. Fritz and W. Fehr, *Handbuch Medien: Computerspiele – Theorie, Forschung, Praxis* (Bonn: Bundeszentrale für Politische Bildung, 1997).

Gieselmann, H. *Der virtuelle Krieg – Zwischen Schein und Wirklichkeit im Computerspiel* (Hannover: Offizin, 2002).

Gieselmann, H. 'Spiel mit dem Terror', *heise news*, 20 August 2002. Available at http://www.heise.de/newsticker/data/hag-20.08.02-000.

Gieselmann, H. 'Spielplatz Zweiter Weltkrieg', *c't*, No. 7 (2003), 94–9.

Gieselmann, H. 'Braune Minderheit', *c't*, No. 8 (2003) p. 48.

Grossman, Lt Col. D. *On Killing: The Psychological Cost of Learning to Kill in War and Society* (Boston: Little, Brown & Co, 1996).

Grossman, Lt Col. D. *Stop Teaching our Kids to Kill* (Boston: Little, Brown & Co, 1999).

Holert, T. and M. Terkessidis, *Entsichert – Krieg als Massenkultur im 21. Jahrhundert* (Cologne: Kiepenheuer & Witsch, 2002).

Leiner, M.K. *Schlachtfelder der elektronischen Wüste – Schwarzkopf, Schwarzenegger, Black Magic Johnson* (Berlin: Merve, 1999).

Mertens, M. and T.O. Meißner, *Wir waren Space Invaders – Geschichten vom Computerspielen* (Frankfurt am Main: Eichborn, 2002).

Meves, H. 'Das falsche Spiel mit der Gewalt – Computerspiele und die Gewalt in der Gesellschaft', *telepolis*, 29 July 2002. Available at http://www.telepolis.de/deutsch/special/game/12973/1.html.

National Research Council et al. (ed.), *Modeling and Simulation: Linking Entertainment and Defense* (Washington, DC: National Academy Press, 1997).

Osunsami, S. 'Simulated Sniping – US Army Recruits Teens With Internet Game', *abcNews*, 31 October 2002. Available at http://abcnews.go.com/sections/wnt/DailyNews/army_game021031.html.

Poole, S. *Trigger Happy: The Inner Life of Videogames* (London: Fourth Estate, 2000).

Rötzer, F. 'Üben für den Krieg im Irak – Wartainment: Computerspiele für den Krieg und zur Anwerbung', *telepolis*, 6 October 2002. Available at http://www.telepolis.de/deutsch/special/game/13367/1.html.

Streibl, R.E. 'Krieg im Computerspiel' (1996). Available at http://www.bpb.de/snp/referate/streibl.htm.

Streibl, R.E. 'Spielend zum Sieg!' (1996). Available at http://www.bpb.de/snp/referate/streibl2.htm.

Thompson, C. 'Violence and the Political Life of Videogames', in L. King (ed.), *Game On: The History and Culture of Videogames*, exhibition catalogue (London: Laurence King, 2002), pp. 22–31.

Villanueva, Lt Col. F. and Maj. A. Huber, 'Out of the Box – Using COTS Products to Build Collective Skills', *Training and Simulation*, June/July (2002), 40–5.

Willmann, T. 'Death's a Game', *telepolis*, 17 June 2002. Available at http://www.telepolis.de/deutsch/kolumnen/wil/12679/1.html.

Willmann, T. 'Ganz anders als Krieg sollte ein gutes Spiel immer Spaß machen', *telepolis*, 26 July 2002. Available at http://www.telepolis.de/deutsch/special/game/12928/1.html.

Woznicki, K. 'Krieg als Massenkultur', *telepolis*, 24 August 2002. Available at http://www.telepolis.de/deutsch/inhalt/co/13059/1.html.

Wright, K. 'Does Media cause Violent Behaviour? A Look at the Research', *womengamers.com*, 6 October 2000. Available at http://www.womengamers.com/articles/gameviolence1.html.

3
Strategic Information Warfare: An Introduction

Gian Piero Siroli

> ...Attaining one hundred victories in one hundred battles is not the pinnacle of excellence. Subjugating the enemy's army without fighting is the true pinnacle of excellence
>
> (Sun Tzu, 'The Art of War', about 500BC)

3.1 Introduction

Since the mid-1980s a very rapid evolution of information and communication technologies (ICT) has taken place, together with a worldwide proliferation of information systems. The rapid expansion and integration of telecommunications technologies, computer systems and information processes has deepened and broadened the Information Infrastructure (II) at every level of society and, in particular, in western industrialized countries; citizens, economic activities and state organizations are increasingly reliant on information technologies (IT).

This evolution process has many positive aspects. However, it should also be analysed from the point of view of an increasing dependence on the new networked global II currently under construction and, as a consequence, also in terms of vulnerability and possible security implications. Widespread reliance on information-based technologies may be driving society towards an unprecedented degree of global connectivity and interdependence. New vulnerabilities, which can be exploited at various different levels, are induced by the convergence and increased overlapping of the traditional critical infrastructures of a country (for instance, vital infrastructures like the energy distribution systems or the emergency services) with present-day II being prone to electronic attacks.

The IIs, consisting of information systems and telecommunication networks with all their related technologies, is also becoming increasingly important for the defence policies of many countries since it might become a significant military target under certain conditions. Moreover, it should not be forgotten that information and disinformation has always been a key factor in war. The exploitation of advanced IT in the military field is driving the development of new warfare techniques, raising the problem of both national and international security.

This chapter is an introduction to the subject of strategic IW – in other words ITs seen in the context of national and international security; it describes possible vulnerabilities of critical infrastructures in modern developed countries. The report released in 1997 by the US President's Commission on Critical Infrastructure Protection (PCCIP) will be taken here as a case study and some of its main conclusions will be analysed and discussed.

What does IW mean? From a general point of view, it includes the actions taken to achieve superiority by affecting an adversary's information, information-based processes, information systems and computer-based networks, while defending one's own domestic II. In other words it is the set of activities intended to deny, corrupt or destroy an adversary's information resources; it includes both offensive and defensive operations, often with a significant overlap between the two.

3.2   Context

The United States of America (USA) is probably the most advanced country in the world in terms of IT. At the same time, it is also the one that is most dependent on communication infrastructures, with the consequence that it is far more vulnerable than other countries with respect to IT. Particularly in the USA, various activities and research programmes concerning IW are taking place at many levels, addressing the questions of protection, assurance and the survivability of vital infrastructure.

These activities include a series of official steps taken by the US government; we will mention here only some of the most important ones to exhibit this trend. In January 1995 the US Secretary of Defense established the Information Warfare Executive Board 'to develop and achieve national IW goals'. Six months later, the Presidential Decision Directive 39 (PDD39) set the policy concerning terrorist threats, which also includes activities relating to IW. In July 1996, the Executive Order 13010 established the President's Commission on Critical Infrastructure Protection (PCCIP), setting the goal of assessing 'physical and cyber threats to national vital infrastructure' and developing 'strategies to protect it'. At the same time the Infrastructure Protection Task Force (IPTF) was created, 'to increase coordination on infrastructure protection'.

The key elements of US policy concerning critical infrastructure protections were defined in the PDD 63, released in May 1998. This directive was followed by the creation of two agencies – the National Infrastructure Protection Centre (NIPC), located at the FBI, and the Critical Infrastructure Assurance Office (CIAO) at the Department of Commerce. At the same time other projects were proposed, for example, the Federal Intrusion Detection Network (FIDNet) to protect government and key private sector nodes through widespread system and network monitoring.

In July 1999 the Executive Order 13130 established the National Infrastructure Assurance Council (NIAC). Later on, in January 2000, the US administration issued a 'National Plan for Information Systems Protection', describing the new dependencies and threats. This proposed a public – private partnership and training programmes to achieve cyber defence; this plan officially included FIDNet for the protection of federal civilian agencies whose funding requests sum up to $10 million in the Fiscal Year 2001. The FIDNet initiative, a warning system to monitor critical computer networks, was later abandoned and replaced. However, the goal is to sustain the ability to alert NIPC in case the Federal Computer Incident Response Capability (FedCIRC, a central coordination and analysis facility dealing with computer security related issues) suspects any hostile activity. In 2002 the US 'President's Critical Infrastructure Protection Board' published a report on 'The National Strategy to Secure Cyberspace' and the Joint Economic Committee of the Congress released 'Security in the Information Age', describing a range of perspectives on infrastructure protection. Since then, many more activities have been developed in this context at many levels.

Similar programmes began later in Europe, albeit with a different aim. In 1997 and 1998 four workshops were held in consultation with industry, academia and public authorities to prepare for the establishment of the European Dependability Initiative within the Information Society Technologies (IST) Programme, to be managed by the Information Society Directorate-General (DG) of the European Commission. The goal was to raise and trust and confidence in systems and services, addressing dependence on ICT and new vulnerabilities.

In 1999 the Scientific and Technological Options Assessment (STOA) team commissioned four studies, in response to a request from the Committee on Citizen's Freedom and Rights, Justice and Home Affairs. The first study concerns state-of-the-art electronic surveillance via Communication Intelligence (COMINT) for global interception capabilities. The second study deals with encryption and cryptosystems, the mechanisms used to protect against interception of communications. The third study examines the legality of intercepting electronic communications, reviewing the existing policies and international agreements. The final study analyses the economic risks arising from the interception of communications. These activities are focused in a slightly different direction than the US initiatives and include data protection and the confidentiality of communications; it is an indication that these technologies have important implications in very different sectors. It is appropriate to mention here that in November 2001 the Council of Europe signed a 'Convention on Cybercrime'; cyber crime is just one more aspect to take into account, even if not one of the most important ones, in relation to national and international security.

In recent years, however, some European countries, including Germany, The Netherlands, Norway, Sweden, Switzerland, and the UK, have started initiatives on vulnerabilities analyses of their infrastructures, a sketch of early warning systems, in some cases also suggesting countermeasures and setting policies within this framework. Austria, Finland, France, and Italy are also becoming increasingly active in this domain.

The United Nations (UN) also recognised the importance of this issue. In December 1998, the General Assembly released Resolution 53/70, addressing the security of global information and telecommunication systems and promoting the consideration of existing and potential threats in the field of information security. In 1999, two UN agencies – the Department of Disarmament Affairs (DDA) and the Institute for Disarmament Research (UNIDIR) – organized a discussion meeting on 'Developments in the field of information and telecommunications in the context of international security'. If we exclude bilateral and multilateral contacts, this was the first meeting on this topic to be held within the UN community. In December 1999, a second Resolution (54/49), collecting the views and assessments of a certain number of countries, invited member states to define basic notions related to information security and the development of international principles in order to enhance the security of global information and telecommunications systems. Since then, further resolutions have been adopted by the General Assembly (55/28, 56/19, 57/53, 58/199), indicating the increasing interest around this topic within the UN.

3.3 Critical infrastructures

What exactly do we mean by the term 'infrastructure' in this context, and why does it need protection? An infrastructure is a framework of interdependent networks and systems, generally interlinked at many different levels, including industries, institutions and distribution capabilities that provide a flow of products or services. Some infrastructures are becoming essential, if they are not already, for the organization, the functionality and economic stability of a modern developed country. To be more specific, it is possible to identify five main sectors (following the scheme of the report entitled 'Critical Foundations', released by the PCCIP Commission), each one including very broad domains:

1. Information and communication.
2. Energy.
3. Banking and finance.
4. Physical distribution.
5. Vital human services.

This section will address each of these items in turn. The 'Critical Foundations' Report will be referred to as the 'PCCIP report'. One should not forget, however, that this is just one of many possible schemes to describe and analyse the complexity of the problem; different approaches are possible, in terms of components, networks, services and domains. These infrastructures are considered 'critical' in the sense that they are supposed to be indispensable for normal day-to-day civil life and their incapacitation or destruction would have a debilitating impact on economic security or the defence capabilities of a country. It is worth pointing out that these five sectors are not independent, but very strongly correlated to one another. What follows is a very concise description of the critical infrastructures included in each of the five sectors.

The *Information and communication* sector includes all the telecommunication equipments, the computer and network technologies and techniques (both hardware and software), the lines providing connectivity and Internet-based services. It includes the Public Switched Telephone Networks (PSTN) providing voice, data, video connectivity and private lines, in addition to the millions of computers used for commercial, academic and government use and in private homes. This sector includes the support for processing, storage and transmission of data and information, including the data and information themselves. Currently, we are witnessing a global merging of all of these infrastructures.

The complex systems of production, storage and distribution of every form of energy characterizes the *Energy* domain: natural gas, crude and refined petroleum, nuclear power, including processing facilities, and electricity. For example, the electrical power grid of a country is part of this infrastructure; this domain also fuels the transportation services, manufacturing operations and home utilities, and is essential to many other infrastructures. It is a key component to other infrastructures and vital for the economic stability of a country.

The *Banking and finance* sector includes entities such as banks, commercial organizations, investment institutions, trading houses and associated operational organizations and support activities like financial transaction services, electronic payments and related messaging systems. To give an example, in the USA this infrastructure manages trillions of dollars – from individual deposits and pay cheques, to transfers for major global enterprises.

The networks of roads and highways, railways and the airspace system (airlines, aircraft and airports) characterize the *Physical distribution* sector, which also includes national pipelines, ports and waterways. This infrastructure allows the movement of goods and people within and beyond the borders of a country.

Finally, the *Vital human services* sector includes emergency services (for example, police, fire-fighting and rescue services), government services, state and local agencies, and country-wide water supply systems serving, among others, agriculture, industries and homes.

The mosaic of interconnectivity makes the global infrastructure extremely complex. It is extremely difficult to define and establish exact boundaries, measure impacts of events and identify clear responsibilities for the management of the different frameworks. It should be noted that two infrastructures – the 'Energy' sector (in particular the distribution of electric power) and the 'Information and communication' sector – underpin the other infrastructures, so that the interruption or disruption of these sectors could potentially have the widest effect. The current trend is that all the critical infrastructures are increasingly dependent on ICTs.

In addition to natural disasters, failures and human misbehaviour, each of these infrastructures, depending on its design, implementation or operation, can be susceptible to destruction or incapacitation and is vulnerable to some extent. This vulnerability can be at the physical level, at the cyber level, or at any combination of the two; this combination, in particular the cyber physical dependence, is the most obscure sector. The problem arising at the beginning of the Year 2000 from the incorrect handling of a two-digit year date format in many application programmes, the well-known 'Y2K bug', can be considered to be an example of this. The attention paid in estimating the possible consequences produced by Y2K, in particular by western countries, shows that already it is difficult to assess the effects of a single software bug, relatively simple and not malicious, distributed over many systems spread over our basic information infrastructures. Even if the Y2K problem might have been overstated for commercial reasons or by the media, it is a fact that users, and sometimes not only end-users but also professionals, are not fully aware of all the low-level detailed features (not to mention real software bugs) within each application and, most importantly, all of the indirect consequences of these 'features', especially in complex systems. This problem is particularly evident in the security domain.

3.4 Vulnerabilities

What are examples of possible vulnerabilities within the various domains? 'Energy' and 'Physical distribution', in particular, may suffer physical vulnerabilities to various degrees, caused, for instance, by natural disasters or sabotage, but here we wish to focus on possible problems and threats of a different nature.

Information and communications: in addition to natural disasters, the primary threats to this sector are system failures and instabilities arising from the increased volume and complexity of interconnections. In the past there have been documented deliberate attacks and intrusions through the software-based disruption of network devices and management systems. In recent years PSTN has become increasingly software driven, remotely maintained and managed through computer networks, which has increased the possibilities of electronic intrusion. The existence of mega-centres for operations support creates single points of failure and makes the targeting of hostile actions easier. The infrastructure vulnerability has probably grown during the 1990s; as far as the Internet is concerned, high-level security was not a primary design consideration during its evolution and deployment.

Energy: the level of vulnerability of this sector has been increased by the recent rapid proliferation of industry-wide information systems based on the open architectures used in the operating environment. This includes increasing reliance on communication links, which sometimes runover public telecommunication networks. As a particular example, the widespread and expanding use of the Supervisory Control and Data Acquisition systems (SCADA) to monitor and control energy infrastructures, runs the risk of serious damage and disruption by cyber means. SCADA is employed by the electric power, oil and gas industries. Possible electronic intrusion through public networks could cause significant disruption if an intruder were able to access the system, modifying the data used for operational decisions or taking control of procedures for critical equipment. Dangers also come from the extended use of commercial off-the-shelf (COTS) hardware and software. COTS are considered risky because detailed specifications might not be available or may simply not be met by some of the components, causing limitation of functionality or faults because of the presence of lower quality standards; they sometimes have built-in vulnerabilities and may pose problems of security and dependability. In addition, sometimes vulnerability information, useful for the targeting of traditional military activities, is made publicly available.

Banking and finance: this is considered the safest domain, and the main vulnerabilities are of a physical nature. Strong measures have been taken, especially in the USA, to harden primary facilities, to secure the infrastructure and to provide extensive system redundancy; however, there remains some level of risk from the disruption of telecommunications and electric power services. In addition to large-scale infrastructure vulnerabilities, this area suffers because of significant opportunities for theft and fraud in individual institutions. Insiders, who might use authorized access to collect confidential information or operate systems for personal profit, constitute the most persistent security threat. Due to its intrinsic sensitivity and in order to maintain public confidence, financial institutions will often refuse to use external agencies in problem reporting, reducing the transparency of the system and making the discovery of intrusions and the protection of the overall infrastructure sometimes more complicated.

Physical distribution: as in other areas, cyber vulnerabilities are emerging, as this sector relies increasingly on IT and communications infrastructures. Every aspect of the transportation industry is affected – for example, the rapidly expanding use of Intelligent Transportation Systems to optimize and increase overall efficiency. In some cases, data publicly available on the Internet could be used to collect information on potential military targets.

The PCCIP report states that, in the USA, the most significant projected vulnerabilities are considered to be those associated with the modernization of the National Airspace System (NAS) for air traffic control. This includes plans to adopt the Global Positioning System (GPS) as the sole basis for radio navigation in the country by 2010. At present, NAS is relatively immune from intrusions, being composed of difficult-to-penetrate dedicated subsystems and networks. The newly planned architecture is likely to use open systems and shared communications networks in conjunction with COTS hardware and software products. As a consequence, the risk of unauthorized access and the probability of malicious actions would increase substantially. As far as GPS is concerned, current plans could lead to overreliance on this system, which is vulnerable to jamming (transmission of noise interfering with original signal) and spoofing (broadcast of false GPS information).

Vital human services: in this sector the main concern in relation to cyber vulnerabilities is probably the increasing reliance on SCADA systems being used for the control of water supplies; in addition, some emergency systems can be overloaded through misuse. Government services keep mega-databases containing highly confidential information on private citizens; cyber intrusion into these databases is a concern as is, once again, the general dependency on computer technology. In addition, cyber reconnaissance to track military assets might be possible in some cases.

More detailed examples can be provided: the first one concerns PSTN, where the level of vulnerability is growing. In recent times, the number of interconnections among telephone companies has increased, including, in particular, interconnections through the Internet. This means that two different telephone networks, using SS7 (Common Channel Signalling System 7, known also as C7) standards, can be interconnected through an Internet Protocol (IP) packet network like the Internet. In other words, a phone call can be transmitted from the caller's local switching point to a 'SS7-IP gateway', travel through an IP network to a second gateway where it re-enters a different telephone network in order to reach its final destination. SS7 is a global open standard defined by the International Telecommunication Union; it describes procedures and protocols by which PSTN network elements exchange information over a digital signalling network for call set-up, routing and control. SS7 was originally designed for a closed community of telephone companies, but recently there has been a proliferation of new services and a significant increase in the number of SS7 vendors providing both hardware and software products. This trend necessarily induces a relatively high level of information sharing and standardization, increasing the overall vulnerability of the global system; many more actors are now present on the scene, a situation that is creating opportunities for insider attacks. The main point to stress here is the relatively recent interconnection between the traditional telephone systems and digital data network: the IP 'trunk' is relatively easier to intercept than the traditional SS7 traffic. In some cases, existing SS7 firewalls might not be adequate or reliable enough, allowing external IP packets, injected into the Internet in the proper format, to enter the telephone network through the SS7–IP gateway. More generally, leaving aside SS7, many present switchboard systems can be remotely managed through their network connections and are built on top of computers running standard operating systems, with known vulnerabilities. Among possible consequences of an intrusion are unauthorized call control or modification of call routing tables within the telephone exchanges.

Another example of vulnerability is the transport architecture of switched networks. As mentioned also by the PCCIP report, many of the fibre optic network installations by commercial carriers are configured as Synchronous Optical NETworks (SONET), a standard for physical-level transport, supporting Asynchronous Transfer Mode (ATM) based services, present at the network backbone level. In SONET most of the elements are remotely managed through packet data network connections, which are somewhat vulnerable to electronic intrusions; in addition, the maintenance and testing ports of network devices could be remotely attacked. Even if it might be less relevant these days because of changing technologies, in the past this was the cause of a large-scale network outage produced by a cyber attack.

One more example concerns emergency systems; in April 2000 NIPC released an alert on a 'Self-Propagating 911 Script', spreading through four of the major US Internet service providers where thousands of computers were scanned for disseminating the malicious script. Victim systems would dial 911, an emergency phone number in the USA, causing authorities to check out substantial numbers of false calls and overloading the infrastructure.

Coordinated Distributed Denial of Service (DDoS) attacks (where servers are flooded by a number of request messages they cannot cope with, originating from multiple locations on the Internet) in some cases produced real and substantial financial loss. This was the case in February 2000, when a number of high-profile attacks temporarily disabled some important electronic commerce Internet websites; sophisticated DDoS tools appear to be undergoing active development, testing and deployment over the net. Recent relatively sophisticated attacks appear to have been planned for weeks or months, since they require clandestine loading of hacking software onto hundreds of computers around the world.

In the previous section SCADA systems have been mentioned; together with DCS (Distributed Control Systems), they are part of the larger class of industrial control systems, often used for operation and maintenance of critical infrastructures. These systems, used for data acquisition (through monitoring sensors) and control (through actuators), perform key functions in providing essential services for electricity generation and distribution, for the water supply infrastructure, waste treatment systems and oil and gas industries. These control networks were initially designed to optimize functionality, but they paid little attention to security which, in many cases, could be considered weak or non-existent; in the past, this was not a problem since systems were completely decoupled from any other network and were basically accessible only by authorized operators on dedicated infrastructures. Basically, old architecture was not designed for the current transition from the 'analogue' to the 'digital' world. More recent control systems using SCADA rely heavily on digital information technologies, using standard software tools, operating systems and communication protocols; in some cases, the control system is interlinked with other general purpose network and is being operated in a way for which it was never designed. As a consequence, new control systems inherit vulnerabilities from the IT sector and become prone to cyber-based attacks. Often there are inadequate password policies, there is no protection against data interception or manipulation, commercial operating systems and communication protocols have known weaknesses and often there is no protection against spoofing in the underlying low-level communications. Some logic controllers could even crash (thereby losing control of the device) under a simple remote port scan; and viruses and worms could probably be specifically designed to target SCADA infrastructures.

3.5   Actors: how and who

The examples in the previous section show how the basic communication infrastructure, including both the telephone and Internet networks, can be vulnerable under certain conditions; it is important to point out that other sectors rely on them for their normal day-to-day activities. All of the critical infrastructures are increasingly interconnected through communication networks. This relatively recent trend increases global efficiency but, as a side-effect, it decreases resilience; it is recognised that mutual dependence and interconnectivity bring new vulnerabilities. The management of complex interconnected systems is a difficult task, especially because of their interdependencies. From the security point of view, the risk lies at different levels, ranging from generic crimes like frauds or criminal activities using the net, to sabotage, interception and intrusion. The spectrum of targets is also very broad, from individuals to institutions. Restricting discussion to the IT sector only, the possibilities of break-ins and hacking are high at the user, computer, and network levels. Various sorts of 'Hacker Kits' are freely available on the Internet; the tools are so numerous and so varied (and often sophisticated) that some kind of zoological approach would be needed to classify them. For example, it is possible to map the network topology using 'scanner' programmes and intercept and look at the content of packets travelling through the data lines using 'sniffer' applications. It is possible to hack or poison the 'Domain Name Service' (DNS, a basic functionality translating IP addresses into computer names) or produce broadcast storms that may drastically reduce network availability. Under certain conditions one can remotely crash or shut down computers or network devices or limit some of their functionality. In order to gain access to computer systems, it is possible to use password cracker programs or exploit 'buffer overflows', executing code in reserved and unprotected memory space. To this list of dangers we can add Trojan horses (disguised malicious applications), viruses (self-reproducing code attached to executable code) or worms (autonomously transferring replicas of themselves over the network). In general, there are two main phases in mounting an IW attack: the first step is to perform a detailed mapping of the net, collecting data on active network devices to carry out a vulnerability analysis. In this respect, many networks worldwide detect a more or less regular activity of mapping, often performed by unidentified sources. In the second phase, the appropriate software weapon is released. Release does not mean activate; the activation can come later, programmed to occur at a certain time, under defined, logical conditions or following a specific command. In some cases a test reaction can be performed in advance, to ascertain the defence capabilities of the attacked system.

Who are the actors involved in such activities? Here again the spectrum is very wide; with no intent of being exhaustive, it is possible to distinguish a few general classes. Media like TV or newspapers often refer to generic 'hackers', who can be professionals or, more often, amateurs or hobbyists, people who like spending nights in front of a computer screen breaking into electronic systems. They often have no explicit malevolent intent, but view their activities as a personal challenge. A second group includes insiders, who are often involved in cases of industrial, economic or corporate espionage; this group is often motivated by money or revenge and can pose a significant threat for organizations. The third group consists of criminals at the individual level or within organizations, targeting, for example, financial information resources. Corporations actively seeking competitors' trade secrets, often using insiders, can fall in this category. Furthermore, there are politically motivated state and non-state groups, ranging from government agencies like intelligence agencies or military units to terrorist groups; their goals can include information collection, propaganda, electronic surveillance, censorship and sabotage.

Concerning the resources required for such an activity, it should be pointed out that even if the entry cost of micro-computing and networking devices is relatively low (a simple, cheap, PC with a modem can be sufficient to annoy system administrators), in order to become a meaningful actor in this context a fair amount of intelligence gathering is needed, together with a high degree of technical expertise and the availability of a large amount of resources.

3.6 Open questions and comments

Before making some general comments, let us refer once again to the PCCIP report of October 1997 and summarize the USA's strategic objectives as set out in this document. This report recognises that the technological dependence on critical infrastructures is increasing and that there is a widespread capability to exploit infrastructure vulnerabilities. It also states that in society, there is insufficient awareness of this topic and it suggests various actions to be taken by the government. First, the problem needs to be defined more precisely. A systematic examination and a very detailed evaluation of critical infrastructures have to be performed in order to propose a precise strategy to protect them. The interconnectivity among different systems has to be analysed in detail, together with the cyber/physical interdependency, to assess the level of vulnerability; the complexity of the problem has to be addressed and the current level of protection and risks understood. The second logical step is to gather information from the government and infrastructure owners and operators, for example, telephone companies and network providers, so that there can be some understanding of who exactly controls which sectors. In addition, it needs to be clarified where responsibilities lie – if they are public, private or shared – in order to understand who is supposed to take action. It suggests starting a close public–private coordination and cooperation between government and industries, promoting a partnership to accomplish their specific infrastructure protection roles. The government should take the leadership in information security management activities, promote inter-agency coordination and integration, and sponsor legislation to develop the legal framework in order to increase the effectiveness of protection efforts. The national awareness of infrastructure vulnerabilities and threats should be raised through education and other appropriate programmes. At the beginning of the Year 2000, the US president began addressing this subject in public speeches. In order to protect infrastructures, the PCCIP report proposed the creation of a national cyber-warning capability, providing immediate real-time detection of attempted cyber attacks on critical infrastructures; the goal is to monitor, provide early warning, alert and respond in order to reconstitute a working minimal infrastructure even with limited functionality. The report also recommends increasing investment in infrastructure assurance research and design (R&D) from \$250 million to \$500 million in 1999, with incremental increases over a five-year period to \$1 billion in 2004. This is an extremely brief summary of the views of the PCCIP Commission; the full document contains very interesting details. Different views exist on the subject, however, the appointed governmental commission made a significant contribution to producing the report and hence it cannot be underestimated.

In January 2000 the White House released the 'National Plan for Information Systems Protection', an attempt to design a way to protect cyberspace;

it can be considered as the evolution of the PCCIP report, which is clearly the starting point. This plan follows very precisely the strategic views of the 1997 report and contains technical R&D and training programmes. In addition, it supports activities to increase public awareness, but also to ensure the protection of civil liberties and protection of proprietary data. A public–private partnership is strongly encouraged to build the base for cyber defence. This plan was supposed to be fully operational by mid-2003; the current US administration seems to share basically the same views on the argument. Apart from the US, a few more countries are currently in the first steps of a detailed analysis of their infrastructures, addressing interdependencies and vulnerabilities.

In spite of all this activity, these are several open questions and some important unresolved issues. In the following, some topics will be briefly discussed, but many more details and analyses can be found in the reference at the end of this chapter, to which the reader is directed. The first issue is about information sharing between public and private sectors: it is evident that in order to reach the goal of centralized analysis and monitoring there is a compelling need for information sharing, up to some non-negligible level, between the two sectors. This sharing can be problematic for various reasons, mainly because of the sensitivity of shared information and the possibly divergent interests of the actors. Security agencies are usually reluctant to release confidential and classified information, while industries in competition in the market would like to retain trade secrets and proprietary information. The situation is made even more complicated if we take into account multinational or foreign corporations. The responsibilities of government and private sector may be conflicting and interests may diverge; an example that occurred in the past was the dispute over encryption between a US citizen and the Department of State. Another topic for debate is the following: what exactly is the government's responsibility? Defence of the country has always been the exclusive preserve of the government, but this may no longer be either true or even feasible, since the civilian sector may no longer be fully protected by interposing military forces. In addition, in the new scenario, private owners and infrastructure operators need to play key roles in infrastructure protection against intrusion, frauds or possible foreign attacks. Where to draw the line between public and private sector responsibility? Can it be drawn at all, or is it becoming fuzzy? A close collaboration between citizens and national security agencies might drive us towards a surveillance society.

The issue becomes more complicated because of the economic deregulation process currently underway, which is causing a higher level of infrastructure vulnerability. The growing fragmentation of systems reduces the control of each individual operator, often limiting redundancy and increasing the overall level of fragility. In the telecommunications sector especially, the appearance of new multiple intermediaries into what were once end-to-end services makes the level of operational interdependence even more complex. As a consequence, the management and coordination of complex systems becomes more and more difficult. The PCCIP report and the national plan suggest and support R&D activities on topics like intrusion monitoring and detection and incident response and recovery. As we mentioned above it also plans to create a national cyber-warning capability. It is interesting here to go into more detail: it implies the capacity for near real-time monitoring of telecommunications infrastructures, the ability to recognise, collect and profile anomalies associated with attacks and, finally, the capability to trace, re-route and isolate the electronic signals associated with an attack. The complexity of the overall system, which is also constantly changing, makes tactical warning and attack assessment an extremely difficult problem. Distinguishing between the 'noise level' of day-to-day accidental events and real attacks and, in addition, being able to trace the source of an attack might be a formidable task. On this subject, it is worth saying that intrusion monitoring applications and products are being built and commercialized by the computer and network industry. The risk of evolving towards a surveillance society might be even higher if legislation is not correctly set up to avoid the misuse of such facilities. In the USA, for example, the conflict could be with the 'Fourth Amendment', protecting individual privacy from unwarranted governmental intrusion. It is appropriate to mention possible legislative conflicts and jurisdictional controversies, a new area on which a public debate would be very interesting. One can ask himself the question whether, in addition to defensive tools and techniques, some actors are actively building offensive info-war capabilities. The boundary between national and international security might become thinner and thinner, like the distinction between military and civil sectors.

3.7 Conclusions

The PCCIP Commission found 'no evidence of an impending cyber attack, which could have a debilitating effect on critical infrastructures' (so no imminent threat had been observed at that time), but a 'widespread capability to exploit infrastructure vulnerability'. As in many other debates, there are radical voices; those who think that hackers are on the verge of destroying basic infrastructures in developed countries. There are also sceptical voices who think that the whole argument is just information mania, and that the catastrophic scenario view is just for 'demo' purposes in order to absorb the vast amount of money made available by the US government to prevent something hypothetical from occurring. The scenario that has been drawn in the previous sections is very complex and this complexity has not been hidden in order to give the reader a feeling of the overall picture. This chapter is intended to be an introduction to the subject so, in order to limit its length, many arguments have been only briefly mentioned and oversimplified, a few have been skipped. In order to disentangle the complexity we will try to focus on some facts in an attempt to summarize the key elements.

Given the recent and ongoing digitalization of industrialized countries, it is evident that critical infrastructures exhibit a growing dependence on networked information systems and communication technologies; this dependence is a source of vulnerability at many levels. Widespread reliance on information-based technologies has resulted in an unprecedented degree of global connectivity and interdependence, making overall management more complex and causing possible disruptions and cascading effects, as in a chain reaction. In addition, from the point of view of security, the super-position and the current process of merging traditional critical infrastructures and information infrastructures increases the global vulnerability, induced by electronic attacks, which was, up to now, limited only to the IT sector.

In the medium term, the development of sophisticated tools and more robust systems might reduce this vulnerability, let's hope this will happen, this is the direction efforts are going. At the moment it is evident that cyber attacks are technically feasible at different levels of complexity. They are not only feasible, but take place all the time in the worldwide networks; the consequences, often not only of economic nature, are difficult to estimate.

It is evident to any computer or network security expert that IT infrastructures can be highly vulnerable – at least locally – to a limited scope attack; small-scale or temporary disruptions can be relatively easy to produce. On a larger scale, it is quite difficult to assess the level of risk and the associated vulnerability. Evaluating the effects of the interconnection and interoperation of very different systems and infrastructures can be an extremely compli-cated task; only now are we learning how to master systems at such a level of complexity.

Possible military and intelligence activities, with both defensive and offensive roles, cannot be excluded in this context; on the contrary, some countries are explicitly addressing them. The exploitation of advanced IT is driving what some people describe as the Revolution in Military Affairs (RMA). Some countries are probably analysing the possible impacts of a potential enemy's information infrastructure disruption. Obviously, national security is of primary importance for every country in the world, but this has to be balanced with the right to privacy and the security of personal and commercial information in a worldwide domain, and, of course, with the national security of other countries. Given the transna-tional nature of the problem, clear international principles need to be estab-lished and agreed upon in order to enhance the security of global information and telecommunications systems. UN resolutions adopted by the General Assembly recognise the need to define notions to deal with unauthorized interference, or the misuse of information systems and resources. For example, appropriate bodies (the United Nations, the International Court of Justice, the G8, bi- and multilateral negotiations or other agencies) might be chosen for dealing with non-peaceful purposes of ICT. The very nature of global networks goes beyond the jurisdictional limits of each country, so an adequate legal framework to develop a uniform legislation is required, as long as there is a clear definition of a chain of responsibilities among the different actors. International cooperation at various levels should be fostered. Some initiatives show that military and civilian activities and functions tend to merge; IT can be considered to be an example of dual-use technology. After all, many scientific and technical developments have had, and continue to have, both civilian and military applications. Cyber development of military affairs is well underway; the 'information weapon' might not be only a virtual concept. The evolution of this field, recognising the broadest positive opportunities, should also be followed in the context of international security, in order to prevent the potential misuse of these technologies for criminal purposes and to avoid undermining international stability.

References

Centre for International Security and Arms Control, *Workshop on Protecting and Assuring Critical National Infrastructure: Next Step* (Stanford, CA: Stanford University, 1998).

Chapman, G. 'National Security and the Internet', paper presented at the Annual Convention of the Internet Society, Geneva, 1998.

Denning, D.E. *Information Warfare and Security* (Boston: Addison Wesley, 1999).

Joint Economic Committee US Congress, 'Security in the Information Age: New Challenges, New Strategies', *Report of the Joint Economic Committee US Congress* (USA, 2002).

McClure, S.J. Scambray and G. Kurtz, *Hacking Exposed* (California: Osborne McGraw-Hill, 1999).

Molander, R. S. Riddile and P. Wilson, *Strategic Information Warfare: A New Face of War* (Washington RAND: MR-661-OSD, 1996).

Neumann, P.G. *Computer Related Risks* (Boston: Addison Wesley Professional, ACM Press, 1995).

Northcutt, S.J. Novak and D. McLachlan, *Network Intrusion Detection: An Analyst's Handbook*, 2nd edn (USA: New Riders Publishing, 2000).

President's Critical Infrastructure Protection Board, The National Strategy to Secure Cyberspace, *Report of the President's Critical Infrastructure Protection Board* (USA, 2002).

Rathmell, A. 'Cyber-Terrorism: The Shape of Future Conflict?', *Royal United Service Institute Journal*, (October 1997).

Siroli, G.P. 'Strategic Information Warfare', *Research Paper 2001/2, Geneva International Peace Research Institute (GIPRI)*, (2001).

Sun Tzu Ping Fa, and R.D. Sawyer, *The Art of War: Sun-Tzu Ping Fa* (Boulder, CO: Westview Press: 1994).

Tanenbaum, A.S. *Computer Networks* (USA: Prentice-Hall International, 1998).

The Report of the President's Commission on Critical Infrastructure Protection, *Critical Foundations: Protecting America's Infrastructures* (October 1997).

The White House, *Defending America's Cyberspace: National Plan for Information Systems Protection* (January 2000).

UN General Assembly Resolutions: A/RES/53/70 1998, A/RES/54/49 1999, A/RES/55/28 2000, A/RES/56/19 2001, A/RES/57/53 2002, A/RES/58/199 2004.

Ware, W.H. *The Cyber-Posture of the National Information Infrastructure* (Washington: RAND, 1998).

Dunn M. and I. Wigert (A. Wenger and J. Metzger (eds), *International CIIP Handbook 2004: An Inventory and Analysis of Protection Policies in Fourteen Countries* (Zurich: Center for Security Studies, 2004).

Part II
Implications of the Problem

4
Virtuous Virtual War
Jari Rantapelkonen

4.1 Introduction

In 2001 General Tommy Franks, commander of the US Central Command, and Mullah Mohammed Omar, Taliban religious leader, agreed that the war against terrorism was not an issue of territory. This view is supported by the fact that the leaders of national security employed John W. Rendon Jr, an advertising professional, to join the ranks of those who are traditionally involved in warfare: politicians, diplomats, soldiers, and reporters. Regarding his role, Mr Rendon has commented:

> I am not a National Security strategist or a military tactician...I am a politician...and a person who uses communication to meet public policy or corporate policy objectives. In fact, I am an information warrior, and a perception manager. This is probably best described in the words of Hunter S. Thompson, when he wrote 'When things turn weird, the weird turn pro'...
>
> If any of you either participated in the liberation of Kuwait City (1991)...or if you watched it on television, you would have seen hundreds of Kuwaitis waving small American flags...Did you ever stop to wonder how the people of Kuwait City, after being held hostage for seven long and painful months, were able to get hand-held American flags? And for that matter, the flags of other coalition countries? Well, you now know the answer. That was one of my jobs. (Miller and Rampton, 2001, p. 11)

'The Rendon Group' was hired by the Pentagon to assist the information-age army by creating a positive image of warfare, thereby helping to win the world wide battle for hearts and minds'. Early in 2002, the group was contracted to work with the Office of Strategic Influence (OSI), a short-lived Pentagon office with a somewhat Orwellian title. The aim of Rendon's public relations firm was to try and create appropriate conditions for the removal of – and encouragement of the overthrow of – Saddam Hussein (Dao and Schmitt, 2002).

This chapter focuses on the relationship between information, information technology, and war. The use of words, images and perceptions in the war against terrorism (September 11, Afghanistan, Iraq) are examined critically by looking at security narrative as a way of determining what is perceived as universal reality. The reality thus produced is a result of using ubiquitous information technology for the purposes of national security. The aim of this chapter[1] is to discuss and attempt to discern and outline the limits of tolerance in information war. The focus is not on looking for the ultimate truth. Rather, the intention of this chapter is to consider the possible results of an uncritical attitude to information (technology) and the arguments of war based on virtuous purposes partly with the help of paradoxes.

War against terror is obviously not a mere information war lacking real threat, suffering, or death. This chapter does not claim that the war against terror could be conceived only as a grand narrative[2] and mental image. Moreover, it is not the intention to imply that the adaptation of the grand narrative of war against terror would not be influenced by actual facts, such as the dictatorship of Saddam Hussein's regime, instead of virtual reality. The aim is, however, to critically assess the narratives that simplify reality, with the assumption that when a state intensively mobilizes its nation and allies – typically supported by the media – the resulting narratives are quite problematic in a variety of ways.

4.2 Theory, information technology and accident

The introduction of two theoretical perspectives is necessary in order to validate the arguments. Firstly, we consider a discussion based on James Der Derian's theory of virtuous war. According to Der Derian (2001), the battles carried out in virtual reality also form the model for real war. At the heart of virtuous war are the superior technical capability and the ethical (good vs bad) imperative to threaten with violence and, if necessary, to actualize violence from a distance.

The model of virtual war gives to war – or in fact returns to it – its virtuous nature. Previously, it was almost impossible to make a distinction between the words *virtual* and *virtuous*, as well as between the two worlds they represent. The arguments of Der Derian (2001) are based on war games as the simulated wars played on computers, wars waged on computer screens do not distinguish between technology (virtual) and ethics (virtuous). Virtuous war creates a paradoxical feeling that war can be bloodless, humanitarian and hygienic.

The model of virtuous war entails that wars are waged the way citizens see them in virtual networks. With the help of high technology, life-like video clips are fed into both television and personal computers. High technology provides a framework for an almost riskless network war. Looking into the core of the virtuous war model reveals a massive force: the Military Industrial Media Entertainment (MIME) network. The wars of our time and age are blended with the wars of the future, which results in disarrangement of reality. CNN news, Hollywood movies, Silicon Valley, and digitized war games paint a very one-sided picture of war.

Simulated war serves to define the practices of real war to soldiers, but is it real? The dream of a perfect war leaves viewers with the illusion that war is virtual, a form of entertainment, a game. In virtual war there is just 'blue', 'red', or 'brown', in other words one and zero, iconic symbols on a computer screen, and therefore not flesh and blood. Der Derian's question concerning information technology is more than topical: will war become easier, peacemaking bloodier?

Secondly, we will employ the theoretical ideas taken from dromology (the theory of speed), developed by Paul Virilio. Virilio has criticized the accelerating speed of modern information and technology. He has indicated that some of the technological developments have spun out of control. For Virilio, information technology, and technological development in general, accelerates events without questioning their contents.

This acceleration has serious consequences. According to Virilio (1986), the new war machine combines a double disappearance; the disappearance of matter and the disappearance of place. Ultimately every technology contains its specific form of accident: the state of emergency with the disappearance of the human race. With information technology abolishing distance, it is possible that what is local becomes blended with the global truth that is representing it.

When adopting ideas of the theory of speed, it is important to take the concept of image seriously because:

> from now on everything passes through the image. The image has priority over the thing, the object, and sometimes even the physically-present being. Just as real time, instantaneousness, has priority over space. Therefore the image is invasive and ubiquitous. Its role is not in the domain of art, the military domain or the technical domain, it is to be everywhere, to be reality. (Virilio, 1988, pp. 4–7)

This leads to the simple idea that virtuality destroys reality, or replaces human reality with technology as *deus ex machina*. Information turns into mere here-and-now communication, the high-speed interconnectivity of which is becoming literally a substitute for the slower-speed intersubjectivity of traditional political systems. This kind of bombardment of information turns mental images, rumours and gossip into a reality that can readily be embraced by those contemplating the background, agenda and ramifications of war. When information is used solely to shed light on a chosen matter, information and disinformation become indistinguishable, since in such circumstances information is not used to convey content based on experience. Using Der Derian's terminology, reality is acquired from the MIME network rather than from personal experience. Therefore, the phenomenology of perception becomes the logistics of perception, in which images war with one another, becoming a substitute for reality itself by means of new technology – the technology of perception.

According to Virilio, new technologies try to make virtual reality more powerful than actual reality. The words and images become more important than the things they represent, which is the true accident. These here-and-now images have priority over their source in information warfare (IW) and they contaminate heterogeneous reality through the homogenization of information, which does not allow any delays. If the same image appears everywhere, it becomes invasive; a grand narrative told by the MIME network with good intentions. It is a form of violence and a threat to democracy because it creates a totalitarian mode of domination. A personal suggestion is that this grand narrative is politically legitimized by referring to its own values, with the narrative paradoxically resulting in an information crisis.

This kind of IW raises the question of an information bomb. Virilio regards information bombs as lethal as nuclear bombs due to their ability to destroy social memory, relations, and international community with an instantaneous overload of unilateral information. This leads to an 'integral accident', in which natural social boundaries are destroyed by global networks, flood of information bypasses deliberation, truth quickly becomes relative, and crises spread like infectious diseases. Local conflicts no longer exist; instead, there is only real-time which does not allow inertia, resistance, or criticism – in the name of effective warfare. Thus, through its ethical imperative, war becomes paradoxically inevitable (Virilio, 1997).

Information technology has altered the perception of warfare. Virilio sees tendencies that have radical consequences, thereby transforming modern/postmodern warfare. Douglas Kellner has argued that Virilio's conception of technology is 'missing the empowering and democratizing aspects of new computer and media technologies'. Kellner's argument is that Virilio's 'vision of technology is overdetermined by his intense focus on war and military technology and that this optic drives him to predominantly technophobic perspectives on technology per se, as well as the new technologies of the contemporary era' (Kellner, 2000, p. 103).

Virilio's theory of speed is worth considering; for example, in modern warfare there is not enough time as human deliberation is minimized. As a consequence of speed, noise and automation, the truth becomes relative. Often, during a military crisis, the media tend to concentrate on specific issues and these issues are intensified through media channels, in contrast to information networks, which offer alternative possibilities. This means that although information technology has the capacity to inform masses of people, it also has the capability to misinform masses of people. This is the issue of information technology and (dis)information warfare that Virilio tries to raise in his discussion. Due to the nature of information technology, like immediacy and ubiquity, different kinds of borders like information and misinformation become indistinguishable. Therefore IW, with the help of information technology, increases the possibilities to distort the reality that Virilio is analysing. This emphasizes the possibilities of usual tendencies of distortion, like simplification that we must be aware of, according to Virilio.

4.3 War on terrorism – the state of emergency

On 11 September 2001, there was more than one attack against the western symbols of power. The media endlessly repeated the acts of terrorism with words and images; in reality there were hundreds, thousands, tens of thousands of terrorist attacks on the World Trade Centre (WTC). Strikes on the WTC and the Pentagon felt impossible to understand, and the media was unable to explain them. It only repeated what had already been seen, as if this very repetition would make the attack easier to understand. The unforeseen series of terrorist acts and the ensuing media spectacle dominated the world by having an effect on people's emotions and behaviour.

Such an example is Annick Wibben, who touches the essence of virtuous war in writing about her experiences on September 11. Pictures shown and seen on computer screens made her feel that she was part of the catastrophe:

> We were all there, but yet we weren't. We saw it, but saw nothing. We kept uttering this isn't real, while knowing that it was. We witnessed death, yet we saw no bodies, no blood. Thanks to information technologies we became participants in 9.11. This confronts us with a new responsibility and a chance of becoming investigators of global terror. (Wibben, 2001)

Terrorism became an easily conceivable and repeatable 'grand narrative', an engine of war. After September 11, there was talk of a 'new war' and 'war against terrorism' following the definitions given by President George W. Bush and Defence Secretary Donald Rumsfeld. It was made very clear to television viewers that the terrorist attacks were an 'act of war' and a 'second Pearl Harbor'. This mental image was strengthened by the fact that the terrorist acts took place on US soil and that the attacks were targeted at the Pentagon, the US military headquarters.

Day after day, the same themes and messages were broadcast and repeated on all television and radio channels, in newspapers, and on the Internet. On computer screens, there was room only for the slogans 'War on America' and 'America's New War'. Once the situation had been defined as war, it was self-evident that there was justification for taking up the challenge with military force. On 7 October 2001, at the commence of Enduring Freedom, the military operation in Afghanistan, the slogan 'War on Terrorism' was accompanied by the words 'America Strikes Back' on the monitors.

The term 'War on Terrorism' was aptly chosen as few people wish the terrorists to win the war besides the terrorists themselves. It is easy to take an oath in the name of war on terrorism, even if there is no agreement about what it means. The answer to the question of what the question was, except a state of emergency in the face of terrorism, was shrouded in mist.

4.4   The necessity and problematics of an enemy

It is difficult for a soldier to go into combat if the enemy has not been defined. According to speeches given by President Bush in September 2001, the US was going to 'eradicate evil from the world' as well as to 'smoke out and pursue evil-doers, those barbaric people', in the war between good and evil. A vague definition such as 'enemy' or 'terrorist' was not sufficient; the enemy was soon given a face, the face of Osama bin Laden and, later, that of Saddam Hussein.

When the enemy is as evil and irrational as the image created, it is impossible to start negotiations. Instead, the enemy needs to be totally defeated and removed from the face of the Earth to keep 'Good' in power. One must ask how that kind of enemy can be defeated in the first place, especially as it has become 'invisible'. Kellner's answer to this is: 'To defeat the bin Laden network thus requires not just the destruction of the Taliban and al-Qaeda group in Afghanistan but an entire global network that will require a multilateral coalition and activity across the legal, judicial, political, military, ideological, and pedagogical fronts' (Kellner, 2002).

Even if capturing the enemy 'dead or alive' was a rapidly established objective, it was a logical one. After bin Laden had been defined as a barbarian, his destiny was not necessarily bound by legislation, but by the rules of the Wild West, which may make more sense to those in the US and Texas than, for example, in Europe. The typical digital nature of information technology (two choices: on or off, 1 or 0) also prevailed in the narratives of political leaders. An either–or situation had emerged (two choices: either you are with us or you are against us) in the battle between 'good' and 'evil' for civilization and against barbarism.

On the other hand, demonizing bin Laden in the process of defining the enemy may also have granted him the status of a hero rising against the West in the Arab world. Still, what can be considered an even greater problem is the ensuing discourse of 'evil', which is based on the binary logic of 'our virtues' and 'their forces of darkness'. The nature of this logic is absolute, since there are no intermediate states between the extremes. It is quite paradoxical that even though this approach allows no room for misgiving, it was the very 'darkness' of the enemy that justified the action that was taken on the basis of mere misgiving. The consistency of this argument was called into question when terrorists were classified as 'unlawful combatants' rather than 'Prisoners of War' (PoWs) when they were taken as detainees to the Camp X-Ray base in Guantanamo, Cuba – even though these citizens of various countries *had* fought in the war.

In the first moments of the war nobody seemed to know the location of bin Laden, the main enemy. It was thought that he might be in Tora Bora, Pakistan, Sudan, or even some where in the USA. A logical conclusion would have been that military attacks should take place everywhere and not only in Afghanistan. Following Kellner's idea, it can be said: to defeat the bin Laden network requires the destruction of an entire network.

The excessive focus on bin Laden meant that the enemy was reduced to an invisible and virtual enemy, someone who was present only in newspapers, on television, on the Internet, and thereby only in our mental images. The sight was set on the face of bin Laden, on the computer screen. The problem with an invisible enemy is that no matter how much military force is used against it, it cannot be destroyed if it remains unseen in the real world. A virtual enemy is thus able to 'move' even into our living rooms. It seems that a propaganda attack against an invisible enemy has an effect only on ourselves rather than on the opponent (Huhtinen and Rantapelkonen, 2002). Adapting and simplifying Kellner's ideas, one has to ask if there really is no other way to win the information war than through the destruction of global (media) networks.

The identification of an enemy is necessary for the legitimization of war. Even if enemies such as bin Laden or Saddam Hussein disappeared or were removed from the computer screens, or they were successfully captured dead or alive, the next 'good evil enemies' would present a new problem. In his 29 January 2002 State of the Union address, President Bush identified the next enemy as the 'axis of evil' (Iran, Iraq and North Korea). In May 2002, this group of 'rogue states' was further extended to include Cuba, Libya and Syria. After Saddam's capture in 2003, new enemies were born.

Destroying the 'axis of evil' will not necessarily be enough, however, as it has been claimed that terrorist groups exist in more than 60 countries. This provides further arguments for the enemy theory, because, according to this theory, a friend turning against us can soon become reality (Harle, 2000). The problem with the 'axis of evil' discourse is that it could further alienate the countries involved and increase the resertment of the western world. In addition to the fact that an enemy can also be a friend, 'the enemy within' is not just a myth.[3]

Technology alone cannot solve the problem. For example, during the 1990s there were stories that Slobodan Milosevic and Osama bin Laden were men that you could do business with. Diplomat Richard Holbrooke pondered later how this ex-friend bin Laden, trained by the CIA, could outcommunicate the world's leading communications society from a cave (Hoffman, 2002). Even though the West controlled the digital information media, the terrorists, with their asymmetric means, were able to create an impact, which the controllers and the messengers of the media were unable to counteract.

Considering the rhetoric problem of the enemy on a practical military level, the fundamental challenge for a soldier, special forces and intelligence and information systems is how to really recognise the enemy, because digital information is not enough in the postmodern battlefield. A soldier has to ask himself how he will be able to tell an enemy soldier from a civilian in Baghdad or, in the case of Afghanistan, how we will be able to tell an al-Qaeda terrorist from a freedom fighter. These groups have made attempts to liberate themselves from each other for decades.

4.5   War on Afghanistan – from postmodern moments to information isolation

Due to information technology, rumours and myths spread quickly in the 'global village'. Military leaders, soldiers, as well as those surfing the information networks cannot verify the truthfulness of these rumours. This is how Howard Kurtz describes the first postmodern moments of the military attacks on Afghanistan launched in October 2001:

> As the cable networks switched to 'America Strikes Back' logos, MSNBC's Brian Williams described some of the warplanes involved by saying, 'If the viewers can picture in their mind's eye ...'

> There were a few minor coups. Jennings, for instance, snared a telephone interview with Abdullah, foreign minister of the Afghan rebels known as the Northern Alliance.

> But for the most part, confusion reigned. CNN showed al-Jazeera footage of a Taliban minister claiming that 'we shot down a plane' during the attacks – an assertion denied by the Pentagon. 'There is no way for us to verify any of this,' Brown said.

> At 5 p.m. on Fox News, anchor Shepard Smith asked reporter Steve Harrigan in Afghanistan whether he knew anything about reports of a second wave of attacks on Kabul. 'No, Shepard,' Harrigan said by phone, 'but you can really sense a buzz of activity here among the opposition' to the Taliban.

> When CBS returned to the Washington airwaves, viewers saw Rather look down and read from a Reuters dispatch. 'Dateline, Kabul,' he said. 'Strong explosions rocked the northern districts of Kabul today ...' (Kurtz, 2001, C1)

The news broadcast during the first few days of the war in Afghanistan consisted of a screen with a few occasional flashes, but was mainly dominated by pictures of snow. Despite this, the reporter, or the expert, gave a convincing account of what was going on in the war. Even now, the events on the battlefield are a matter of guesswork to most of us.

In IW it is important, however, that events, texts, and images have not just been harmonized but that there is also an effort to get approval for them (Luostarinen, 2002). The Joint Chiefs of Staff have confirmed this by stating that information operations aim at systematically moulding the attitudes and emotions of its subjects favourable to the US goals (Joint Chiefs of Staff, 1998, II-1–II-7). According to President George W. Bush speech on 13 February 2001, 'the best way to keep the peace is to redefine war on our terms'. In the beginning of a vague and fragmented war, the media typically seizes the rational grand narratives of national leaders and military command, which are meant to serve a virtuous purpose. The people who produce these narratives and those who recite them and listen to them are usually far removed from the battlefield.

Controlling the environment has always been part of the culture of national security and the military. In this case, the information environment serves the same function as any other battle space; it is important to provide information consistent with the grand narrative, because a chaotic battle space with fragmented, postmodern information is not virtuous.

It has to be noted that the citizens also support the efforts to create a grand narrative. At the beginning of the war on terrorism in Afghanistan, it was suspected that the strategy of information control and restrictions announced by the authorities might lessen the support of the people. According to the polls carried out at that time, the majority of Americans believed that the authorities did not tell them everything that was known before September 11, and that they were not giving all the information they had about the operations against terrorism. Most of the citizens found this acceptable.

Although 'Operation Enduring Freedom' easily won the support of the international and national communities, the starting point for the information campaign over hearts and minds aiming at strategic influence was an extreme challenge for the USA. The USA had fought for its credibility in the Arab world for a long time, with poor results. The USA was seen as evil and as the cause of the problems in the Middle East. The information space of Afghanistan was isolated from outsiders. The Taliban had executed a strategy of information isolation, and thus acquired information advantage in Afghanistan.

Next, brief examples will be provided of the technique of isolating the enemy in order to prevent it from providing the global information village with 'other' kinds of information. These examples pertain to the military level or the area of operation. After the military operations in Afghanistan began in 2001, we were shown images of the power and virtuosity of the armed forces. Firstly, images of precision bombings from video clips were distributed to reporters and to the world in press conferences. Secondly, the armed forces dropped leaflets in the operation area. The third strategy was to drop food supplies from aeroplanes. The armed forces had been harnessed to serve in 'humanitarian' operations, in which the war appeared as virtuous; after all, the idea was to uproot terrorism, destroy 'evil', and achieve peace.

However, these virtual PowerPoint briefings did not really reflect the people involved – or the shocking reality of war. It is for this reason that the virtuous virtual war with precision-guided bombing raises great hopes in the public. Fulfilling these hopes is not possible if the information space is filled with contrary information. Many cases of collateral damage reached the news: bombing the procession of tribal chiefs, a wedding party and the stocks of the Red Cross, killing women and children. In spite of all this, the attacks were defined as 'justified'. Moreover, years after the beginning of the bombings the struggle between and within the tribes still continues, and their opium plantations are in full operation. These images, not fitting the grand narrative, are avoided.

The aim of the Taliban was different, however. The press conferences held by the Taliban were seen across the monitors of the 'global village'. The Taliban claims to be peaceful and unwilling to harm anyone. It has also claimed that bin Laden was not responsible for the terrorist attacks. According to the Taliban, the USA was a terrorist when it kept its troops on foreign territory. Of course, most of the claims of the Taliban were false. Reality and life among ordinary Afghans was terrifying.

However, the fear for the possible information superiority of the Taliban and bin Laden led directly to the destruction of the Taliban's radio and television stations in Afghanistan. The Taliban was prevented from expressing its views to the world by closing its embassy in Pakistan. This also prevented the Taliban from holding press conferences in which the media forces of the 'global village' would have been present. It is a myth that the Taliban and al-Qaeda are mere military adversaries.

4.6   Battle for the strategic truth

The aim of the following discussion is to provide another example of the practice and technique of controversial IW for strategic truth. Despite the challenging odds, the USA was successful in the first phase of the information war on terrorism. This success was based partly on the well-chosen words and images of the leaders of national security and partly on the campaigns that aimed to improve the negative public image with the help of news agencies, PR companies and Hollywood. On the international political front, the success manifested itself in the notion that the war would not be short and that no fast results were to be expected of it. Moreover, the opponent could not create a credible opposition between Christians and

Muslims, even though the PoWs were mainly Muslims. The fact that the women's position in Afghanistan was raised to the fore just before the war also aroused people's sympathies for the war in Afghanistan, even though the problem itself was by no means new.

However, with his videos Osama bin Laden managed to arouse suspicion on the rationality of western military operations. On 10 October 2001, National Security Advisor Condoleezza Rice stated to the media that bin Laden's videotapes should not be shown because they might contain encoded messages or provide bin Laden with a propaganda platform (Fleischer 2001). This caused speculation about censorship, and raised the question of hidden meanings more generally. Nevertheless, the video could be seen on foreign news channels and on the Internet. The aim of bin Laden's videos was to bring about a US retreat from Saudi Arabia and an Israeli retreat from Palestine.

Even though bin Laden's videos were produced for propaganda purposes, the tape found in December 2001 in Jalalabad, in which bin Laden flaunted his success, was presented as substantial evidence. Over 60 videos found in the summer of 2002 were presented on CNN as proof of the evilness of al-Qaeda. There is a chance that the endless contemplation on whether bin Laden is dead or alive, and whether the tapes are genuine or fake, may serve only to turn the terrorist leader into a myth – defeating a myth is significantly more difficult than defeating an ordinary, living enemy. This might lead to the notion that the coalition is not able to find bin Laden, which would mean that it is not concentrating on defeating the real enemy anymore but just war.

The battle for the images and strategic truth accelerated. According to a *New York Times* article on 19 February 2002, the Pentagon had established an Office of Strategic Influence (OSI). Its aim was to systematically spread information favourable to the operations of the USA. There is nothing new to this virtuous purpose; deception is an integral and natural part of warfare. It was also announced, however, that the aim was to spread stories, information and disinformation to the allies via the Internet and e-mail in such a way that their origin could not be traced to the Pentagon. Journalists, citizens and allies found such targeting of one's own people unheard of. What was peculiar about the OSI case was that a self-evident principle of war was made transparent by the authorities in this way. However, the OSI as an instrument caused such commotion so quickly that a week later Secretary of Defence Donald Rumsfeld announced that President George W. Bush had decided to close the office. Indeed, on 25 February 2002 Bush promised the journalists: 'We'll tell the American people the truth.'

4.7 War on Iraq – differences in perceptions

In September 2002, global perceptions of war moved from Afghanistan to Iraq as President Bush began to request an international legitimization for war from the United Nations (UN). This also meant that the enemy changed from bin Laden to Saddam Hussein in the eyes of the public. With the unanimous approval of Resolution 1441, the UN Security Council authorized the complete and rapid disarming of Iraq's weapons of mass destruction under the surveillance of UN weapons inspectors.

In early February 2003, Secretary of State Colin Powell presented information to the UN Security Council, with the aim of making an unanswerable case that Saddam Hussein had intentions to produce weapons of mass destruction, something which UN weapons inspectors had found no evidence of. In his presentation, Powell used a number of persuasion techniques: tapes of intercepted telephone conversations and testimonies of Iraqi defectors, spy satellite photographs, and PowerPoint slides with details of secret Iraqi weapons programmes and links between terrorist networks and Baghdad. According to the Americans, it was 'compelling evidence'; still, it was not enough for a second UN Security Council resolution which would have authorized military action (Walkom, 2003).

The information battle of facts and fiction became harder with the publishing of at least three notable intelligence documents by the British authorities. These documents aimed to show that there is 'a huge infrastructure of deception and concealment' with weapons of mass destruction in Iraq. Shortly after Powell had praised the report as excellent and even cited it extensively at the UN, the British government found itself in trouble after admitting the report was partly a copy. In this report on Iraq, based on 'new intelligence information', 'an inkling of unfounded exaggerations were inserted to strengthen the claims' reproducing even the spelling errors from the thesis of a 29-year-old Californian postgraduate whose study focused on evidence from the invasion of Kuwait 13 years ago. Regardless of these accusations, a spokesman for Prime Minister Tony Blair considered the report 'solid' and 'precise' (Hinsliff et al., 2003). These weighty arguments for war against Iraq got into the middle of crises of information, crises of legitimacy.

Arguments for war were virtuous. President Bush, for instance, said in his radio address on 1 March 2003: 'We also stand for the advance of freedom and opportunity and hope. The lives and freedom of the Iraqi people matter little to Saddam Hussein, but they matter greatly to us.' The USA was not just eliminating a threat but delivering a promise of democracy to a region steeped in tyranny.

Despite the presence of this kind of grand narrative of freedom, many important countries – such as China, France and Germany – remained suspicious of these motives. They also had doubts about the the quality of the evidence provided and were frustrated by US arrogance. Arms inspections were still well suited for their purpose. Johanna McGeary writes in *Time*: 'To many Europeans, this war looks like U.S. imperialism. And hypocrisy: they don't see why diplomacy can deal with North Korea's nuclear weapons program but not with Iraq's, or why UN resolutions should be enforced on Iraq but not on Israel.' Also, Russian Foreign Minister Igor Ivanov stated that the US 'must be careful not to take unilateral steps that might threaten the unity of the entire [anti] terrorism coalition.' (McGeary, 2003).

Virtual war had been waged for months before missiles were actually launched towards Baghdad. Different alternatives and models for the attack were simulated on various Internet pages. At the same time, radar stations in Iraq were being bombed in reality. Thus, the war against Iraq was simultaneously both waged and not waged. *The Washington Post* outlined the military situation just before the rhetorical declaration of war on Iraq:

> First phase is already underway. Special Operations troops are executing missions inside Iraq to prepare the way for later attacks. U.S. and British warplanes ostensibly enforcing the 'no-fly' zones in northern and southern Iraq have increased the number and intensity of airstrikes, and recently expanded their list of targets to include Iraqi surface-to-surface missiles… 'We've already got a lot of stuff underway – the air campaign, psychological operations, Special Ops,' said Robert Andrews, a former Pentagon official who oversaw Special Operations activities. (Ricks, 2003, A01)

Real war against Iraq, the pre-emptive one started by the coalition forces led by the US on 20 March 2003, began with 'surgical strikes' aimed at Saddam Hussein and the inner circle of his regime. Then, *déjà vu*, the screens were again filled with crepuscular strikes. Day after day, the same themes and messages were broadcast and repeated all over again on all perception screens; television, radio, newspapers, the Internet, and mobile services. On the real-time screens of the mass media, there was now room only for the slogans 'Operation Iraqi Freedom' and 'War in Iraq'.

Even the same questions were repeated. 'Saddam: Dead or Alive?' 'Is the uniform-wearing man giving a pre-recorded speech on Iraqi state television one of Saddam's body doubles?' 'How many body doubles does Saddam have?' It is well known that rumours spread in such moments when there is no knowledge but only information. It was too early to speculate on finding Saddam's body after a surgical strike and taking DNA samples. Ironically, while the surgery was not successful the patient survived. The impression of *déjà vu* was reinforced by pictures of detainees being treated like PoWs but without an official status. After a later hearing, it will be decided whether they will be released, held as PoWs, or declared illegal combatants. There was nothing new about changing a tyrannical leadership and regime.

Operation Iraqi Freedom was a fact when it popped up and stayed on the screens as a banner. The eight military objectives were outlined on 21 March by the secretary of defense, Donald Rumsfeld. The first objective was, as in Afghanistan, to end (Saddam Hussein's) regime of tyranny, the second was to identify, isolate, and eliminate Iraq's weapons of mass destruction, and the third was to search for, to capture and drive out terrorists from that country. However, the term 'Operation Iraqi Freedom' was problematic. First of all, liberating a nation takes a long time if freedom is connected to democracy. Secondly, it does not sound a bell of war. In contrast to 'Desert Storm', there was no reference to the challenges of actual warfighting.

Stories of a 'cakewalk' and accounts of how Iraqi units would quickly surrender, and how Iraqi citizens would hail the advancing Americans and British as liberators were widespread. There was a will to wage and win the war even before it was started. This psychological effect was strengthened by the overwhelming might of the US military. Also, focusing on the 'Shock and Awe' concept in virtual networks just before the war, created the impression that the campaign would commence with the unleashing of 3,000 precision-guided bombs and missiles. However, the start of the war was disappointing; militarily, it was a smaller operation (under 50 missiles) than President Bill Clinton's use of force against terrorists in Afghanistan and Sudan. Regarding perception management, the public expectations were too high due to the illusion of a war without a war, or at least of a clean war.

Media was excited by the first hits, blows and strikes. Veteran reporter Peter Arnett[4] exclaimed: 'An amazing sight, just like out of an action movie, but this is real'. Erik Sorenson, the president of MSNBC, said that 'the technology – the military's and the media's – has exploded'. He likened the change to 'the difference between Atari and PlayStation', adding that 'this may be one time where the sequel is more compelling than the original' (Kakutani, 2003, E1).

Television news coverage turned viewers into 24-hour couch voyeurs. Dedicated 'bloggers' even provided minute-by-minute information updates. This kind of 'progress' was like watching a real movie without having to go to the movie theatre, with the line between reality and fiction blurring still further. Newscasters repeatedly reminded their audiences that the live scenes of the war they were watching were not a movie. Was this impulse so potent that it was able to blur the line between the real and the fictional? Most likely, since even Colin Powell gave the warning that 'this isn't a videogame, it's a war. It's a real war' (Powell, 2003a). He was paradoxically right because the enemy in the battlefield of Iraq turned out to be different from the one in war games.

The information that started to invade the screens was different from what was publicly expected. The first phase of the war turned out to be longer, more brutal, tiring, frightening, and full of problems and surprises. The headlines featured helicopters being shot down, a precision bomb hitting a civilian bus, enemy soldiers faking surrender, PoWs being captured and tortured, deadly friendly-fire incidents, and women and children being shot at. There were no signs at all of weapons of mass destruction or even of the enemy's sophisticated weapons in the first days of the war, yet the enemy did shoot back. The situation in the battlefield was stressful, with numerous false alarms.[5] Even soldiers in the battlefield were still having thoughts about war events as they were just like in a movie.

The coalition of the willing was successful, with fast movements on the ground. It wanted to see different kinds of information being broadcast from the battlefield to the public. This included accounts of the war going according to plan, a heroic rescue operation of a PoW, and the efficient and clean elimination of military targets. One has to ask if these are more real facts of war than those mentioned in the previous paragraph. What is certain is the uncertainty of war and the uncertainty of war information, even when the best technology of war is in use.

War took a shape that was not equivalent to reality, projecting an image that the public eye did not want to see, a form of war that was resisted. It is well known that virtual war is enough for us common spectators of war. Here is an excerpt from the Central Command briefing on 29 March announced by General Vincent K. Brooks:

What I'd like to show you next is a before and after image set that shows the regime-controlled television studio and broadcast facility. This, like other facilities, was used as part of the command and control network. There were three key facilities that were targeted within it. The post-strike image shows the intended damage at the three arrows. I particularly highlight the topmost arrow, which shows a building that was caved in. Different effects for each weapon system delivered in that complex. (Brooks and Renuart, 2003)

However, the newspapers of the very same day told a different kind of a story about another event of the war with heavy headlines:

At least 50 civilians are believed to have been killed during an air raid on a Baghdad market...Graphic television pictures showed people scrabbling through rubble to reach the dead and injured amid the wreckage of al-Nasser market...Correspondents in Baghdad say there is no clear information yet on what may have caused the destruction of the market...Most of the ground was covered by blood...Central Command... suggested it may have been a misfired Iraqi missile. (BBC, 2003)

General Brooks is more than right in saying that 'the power of information has been the key throughout this operation'. Words by General Charles Horner presented by General Richard Myers on 1 April, 'it depends on what perception is', remind us of the key question: whose information is more real or more valuable? These perceptions of the first days of the war on Iraq speak of the relativity of war. Seen from the village of Umm Qasr, the war appears different than from a London suburb, different from the US Central Command headquarters than from Washington or Europe, different from behind a screen than from within a fighting group even if there were embedded journalists on location.

The US-led coalition of the willing attained an overwhelming military victory. Security is odd, however, because it is a feeling. When the military victory was sealed on 16 April 2003 by General Tommy Franks and his closest officer in a symbolic place, one of Saddam Hussein's palaces, the media were still there. As more and more journalists were heading back home, Walid al-Fartousi, a 33-year-old fruit and vegetable vendor in Baghdad, said that people were beginning to lose their trust.

> America promised Iraq to remove the tyrant government, but now things are even worse. Some people are even beginning to wish Saddam had stayed because all the troubles erupted after his departure ... Until today we are sitting in our houses ... not safe from killers, looters. American forces stand by and do nothing. There is no security, no order. People do not feel safe. (Gordon and Kifner, 2003)

President Bush declared on 1 May 2003, after the successful war effort, that major combat operations were over. Since then, many paradoxical things have happened that have challenged this assertion: not just thousands of local people in Iraq, also more Americans have died than in the first phase of 'major combat operations'. Then the scandal in Saddam's horror-chamber, Abu Ghraib prison, took place, but this time only after the information technology was able to break into the prison in the form of digital cameras. These pictures did not increase the understanding of the nature of warfare. War has also moved to the mosques of Mosul, Fallujah and Sadr City, it had even been said, that the war on terror is not about religion. We have seen through the networks that the US-led coalition is destroying holy places. We have not been aware as much about the reason behind this image, which is 'that everyone of the 77 mosques encountered by Iraqi and coalition forces in Fallujah was used as a weapons storage or a fortress from which to launch attacks' (*Washington Times*, 2004, p. 20).

In general terms, the 'War on Iraq' can be defined as a withdrawal from facts-based pre-emptive war, to the faith-based IW. Another tendency, that has connections to the characteristics of media and speed, is that much of the good work done in Iraq will never be reported, precisely because no news is good news for much of the media, and partly because there is no time for it. Therefore, as the 'War on Iraq' has showed, reality is replaced by fast and simplified news, and speed, without seeing both sides. Therefore information technology has not helped us to be wiser. In fact, as technology and the media have the ability to make sense of our universe, it is becoming less and less grounded in the terrain of experience. This leads Aki Huhtinen to a conclusion that 'manipulation of the symbolic environment can itself produce major events in the political life and the warfare' (Huhtinen, 2004, p. 104).

Afterwards, many people changed their attitudes about going to war. This reflects the fact that we are living in paradoxical times, judging the justness of pre-emptive wars afterwards depending on their success. Judging takes more than here-and-now time, however. It is not only philosophical to ask whether it is necessary anymore to make a difference between real and virtual information to justify war if it is successful. It is interesting to observe how the morally responsible US finds a way to fulfil its virtuous promise to achieve real democratic peace.

4.8 The fog of peace

Virtuous virtual war is not a simple phenomenon. Freedom, democracy, and peace with 'evil' justify the use of force. It is obvious that everyone agrees on the principle of the 'Shock and Awe' strategy, the idea that 'we want them to quit, not to fight'. It is a typical virtuous aim, the pursuit of a general good. Those using force in a war should consider how to use that force in such a way that the statement 'we will not have one bin Laden, we will have hundreds of bin Ladens' in regard to the war in Iraq will not come true.

The one using force is returning to Machiavellian politics, which defines itself not by its excellence but by its outcome. War can be virtuous only if it is effective. Its success will depend more on how pre-emptive the defeat of the enemy is rather than on the extent of its accordance with international law. In this sense, virtuous intentions play a more important role in IW. But virtuous virtual war waged on behalf of freedom, justice, and democracy may leave the spectators with a feeling that those preaching these values no longer believe in them themselves, if they fail to uproot the seeds of terrorism early enough and keep ignoring international conventions. Pursuing an outstanding virtue can be dangerous.

The grand narrative of virtuous war is deceptive if narratives are compressed to form a homogenous story. As such they total words and images into a unity whilst, in such a moment, this virtual unity is more chaotic than ever before in different local realities, as in the battlefield. Facts are not enough in war, however; the will to fight needs to be maintained and, in the name of effective warfare, contradictory opinions are not allowed to be voiced at the time of war. The language of virtuous war is a powerful weapon that can more easily lead us into the wars we should not fight. Fog of information war is thickening due to the fact that there is not one logocentric grand narrative anywhere other than on the screen.

Virtuality has been enabled by technology. According to Virilio's concept of accident, information technology mixes virtual and real, enemy and friend, distant and close, peace and war. A conflict is no longer local, except for the 'no-information-not-connected world'. Virtuous virtual war is being waged all the time all over the world with the assistance of information technology. This can be defined as an integral accident, especially if virtuous and virtual aspects are combined. It is not possible to reinforce 'reality' by reinforcing technology because war is filled with human elements and uncertainty. When speed becomes power in information war, there is no more time left for human deliberation, and therefore speed itself can no longer be called a virtue. What remains in the virtuous virtual war among us, is paradoxical PowerPoint slides free from real, local knowledge. It should be waged, but it cannot be. Dictators should be removed from power, but not with arms. Weapons of mass destruction should be destroyed, but cannot be found. A critical question of enemy should be made: how is it possible that this kind of enemy has emerged, and will killing the enemy on the screen be enough? The conundrum of virtuous war is: 'The more virtuous the intention, the more we must virtualise the enemy, until all that is left as the last man is the criminalized demon' (Der Derian, 2001, p. 101). It seems that war is no longer a social struggle. Cultural borders will very likely turn out to be a gap that is impossible to overcome by means of coercive information. War cannot be won even with high-tech weapons if the enemy is invisible, hiding in society, hiding in the structures.

Acceleration of events and the continuous universal information bombing with video clips and images have changed the nature of war. We can see war, but at the same time we cannot understand it. Naturally, western military and industry deserve credit for 'precision warfare', but, with the help of the media and the entertainment industry, public expectations about this kind of warfare, the RMA utopia, can steer towards misleading perceptions. One of the reasons for this might be that images of war are more attractive, impressive, and persuasive than reality itself. With the 'collateral damages' and 'bomb errors' that the narrative of virtuous war contains, we have ended up in the middle of the reality of war error. This is a disservice to politicians, the military and citizens.

The language that is used in virtuous virtual war loses its meaning if the reality presented in virtual networks becomes a substitute for reality itself. The MIME network and its individual participants such as John Rendon, the advertising professional, and other advertising companies face a considerable challenge when trying to show war in the best possible light, representing reality in a way that Virilio calls 'virtual theatricalization of real world'. The *New York Times* columnist Bill Keller writes about the consequences of war, technology and media: 'Most people did not imagine themselves anywhere near the front line in 1991. Now the front line is where we live, and we are afraid' (Keller, 2003, A17). Waging virtuous virtual war is a double-edged sword. The risk is that whereas the helmet of an American soldier fighting in Vietnam bore the slogan 'War Is Hell', it is possible that in the information war against terrorism, the 'War Is Peace' slogan of virtuous virtual war may be reflected in our images.

Waging modern war in the information age might result in crises of information. The search for the perfect war in the name of virtuous aims produces an illusion of war with the help of virtual technology. This vast information bombing does not help us to look beyond the screens and understand the phenomenon of war. When war is waged for virtuous purposes nobody wants to see the horrific local reality of war beyond the entertaining virtual spectacle. It is easy to understand that 'fog' is the word of wars. This also holds true for wars other than the war on terrorism, war on Afghanistan and war on Iraq. Congo is the latest example of the world's most vicious wars, but we cannot see it in these paradoxical times of information and technology.[6] If we only deal with unilateral information overflow without real knowledge of locality, it can easily make us spectators of the accident, prisoners of permanent war.

Notes

1. I would like to thank Dr Arto Nokkala and Dr Kari Laitinen of the University of Tampere for their helpful comments. This chapter is based on thoughts published earlier in Finland in a multidisciplinary journal *FUTURA* 4/2002. I have revised the chapter thoroughly and updated it taking into account the Iraq situation and the scope of this publication.
2. The term 'grand narrative' was introduced by Jean-François Lyotard. It describes the kind of story that underlies, gives legitimacy to and explains the particular choices a culture prescribes as a possible course of action. Examples of such narratives are Christianity, the Enlightenment, Capitalism, and Marxism. In this chapter I apply Lyotard's concept of a grand narrative to a more limited concept – the 'War Against Terrorism'.
3. This view was confirmed when American Taliban John Walker and the developer of the 'dirty bomb', Jose Padilla, also known as Abdullah al-Mujahiri, were arrested. Another case close to the definition of an 'enemy within' was when Americans were terrorized by Lee Boyd Malvo and John Allen Muhammed. They have admitted that they shot and killed a number of people last fall in the US in the so-called 'sniper case'. In February 2003, catching even a single person (Mr Mohammed) who had represented a grave threat to the US was regarded as a victory by the state. During the Cold War things were different.
4. Arnett was fired from the NBC television company in March, after granting an interview to the Iraqi television saying that the original military plan of the US-led coalition had failed.
5. One of the alarms among American troops in early April informed that Iraq had used chemical weapons. The alarm was false. American troops had been digging the ground in the distance of about 20 kilometres. The diggers had hit a container which leaked material which was initially thought to be a nerve agent, but which later turned out to be harmless industrial chemical.

6. According to the International Rescue Committee report published in April 2003 more than 3.3 million people have died prematurely in Congo since war broke out in 1998.

References

BBC, 'Many dead in Baghdad blast', *BBC News*, available on http://news.bbc.co.uk/2/hi/middle_east/2897117.stm (29 March 2003).

Bush, G.W. 'Speeches and Transcripts'. Available on www.whitehouse.gov (2001–2003).

Brooks, V. and V. Renuart, 'CENTCOM Operation Iraqi Freedom Briefing', *CENTCOM homepage*. Available at www.centcom.mil, presented by Maj. Gen. Victor Renuart and Bris. Gen. Vincent Brooks, 29 March 2003.

Clark, W.K. *Waging Modern War: Bosnia, Kosovo, and the Future of Combat* (New York: Public Affairs, 2001).

Dao, J. and E. Schmitt, 'Pentagon Readies Efforts to Sway Sentiment Abroad', *The New York Times*, 19 February 2002, A1.

Der Derian, J. *Virtuous War: Mapping the Military–Industrial–Media–Entertainment Network* (Boulder, CO: Westview Press, 2001).

Fleischer, A. 'Press Briefing by Ari Fleischer', *Whitehouse*, 10 October 2001.

Gordon, M.R. and J. Kifner, 'U.S. Generals Meet in Palace, Sealing Victory', *The New York Times*, 17 April 2003.

Harle, V. *The Enemy with a Thousand Faces: The Tradition of the Other in Western Political Thought and History* (Westport, CT: Praeger, 2000).

Hinsliff, G. M. Bright, P. Beaumont and E. Vulliamy, 'First Casualties in the propaganda firefight', *The Observer*. Available on http://www.observer.co.uk/waronterrorism/story/0,1373,892146,00.html (9 February 2003).

Hoffman, D. 'Beyond Public Diplomacy', *Foreign Affairs*, 81(2) (2002), 83–95.

Huhtinen, A. 'Soldiership without Existence – The Changing Socio-Psychological Culture and Environment of Military Decision-Makers', in J. Toiskallio (ed.), *Identity, Ethics, and Soldiership* (Helsinki: National Defence College, Department of Education. ACIE Publications, No. 1, 2004), pp. 74–104.

Huhtinen, A. and J. Rantapelkonen, 'Perception Management in the Art of War – A Review of Finnish War Propaganda and Present-Day Information Warfare', *Journal of Information Warfare*, 2(1) (2002), 50–8.

Huhtinen, A. and J. Rantapelkonen, *Imagewars: Beyond the Mask of Information War* (Saarijärvi: Marshal of Finland Mannerheim's War Studies Fund and Finnish Army Signals School, 2001).

Joint Chiefs of Staff, *Joint Doctrine for Information Operations*, Joint Pub 3–13 US Department of Defence, 9 October 1998.

Kakutani, M. 'Shock, Awe and Razmatazz in the Sequel', *The New York Times*, 25 March 2003, E1.

Kaplan, R.D. *Warrior Politics: WhyLeadership Demands a Pagan Ethos* (New York: Vintage Books, 2003).

Keller, B. 'Fear on the Home Front', *The New York Times*, 22 February 2003, A17.

Kellner, D. *September 11, Terror War, and the New Barbarism* (References to the 10 June 2002 version while work was-in-progress on-line at http://www.gseis.ucla.edu/faculty/kellner/kellner.html Published 2003 as *From 9/11 to Terror War: The Dangers of the Bush Legacy* (Maryland: Rowman & Littlefield Publishers, Inc.)).

Kellner, D. 'Virilio, War, and Technology. Some Critical Reflections', in John Armitage (ed.), *Paul Virilio: From Modernism to Hypermodernism and Beyond* (London: Sage 2000), pp. 103–25.

Khalilzad, Z. and J.P. Whiten (eds), *Strategic Appraisal: The Changing Role of Information Warfare* (Santa Monica, CA: RAND, 1999).

Kurtz, H. 'The Fog of War: From the Ground Zero. A Spectral Patchwork Of Sound and Fury', *Washington Post*, 8 October 2001, C1.

Luostarinen, H. 'Propaganda Analysis', in W. Kempf and H. Luostarinen (eds), *Journalism and the New World Order: Studying War and the Media.*, vol II (Gotherburg: Nordicom, 2002), pp. 17–38.

Lyotard, J.-F. *The Postmodern Condition: A Report on Knowledge* (Minneapolis: University of Minnesota Press, Theory and History of Literature, vol. 10, 1984).

McGeary, J. '6 Reasons Why So Many allies Want Bush To Slow Down', *Time*, 26 January 2003.

Miller, L. and S. Rampton, 'The Pentagon's Information Warrior: Rendon to the Rescue', *PR Watch*, 8(4) (2001, Fourth Quarter), 11–12.

Myers, R.B. 'Speeches and transcripts', available on www.defenselink.mil (2003).

Powell, C. 'Transcript: Secretary of State Colin Powell', *Fox News*, available on www.foxnews.com/story/0,2933,82037,00.html (24 March 2003a).

Powell, C. 'Speeches and Transcripts'. Available on www.state.gov (2003b).

Ricks, T.E. 'War Plan for Iraq Largely in Place. Quick, Simultaneous Attacks on Ground and From Air Envisioned', *Washington Post*, 2 March 2003, A01.

Rumsfeld, D. 'Speeches and Transcripts'. Available on www.defenselink.mil (2003).

Susser, E.S., D.B. Herman and B. Aaron, 'Combating the Terror of Terrorism', *Scientific American*, August 2002, 54–61.

Virilio, P. 'Paul Virilio', Interview, *Block*, 14 (1988), 4–7.

Virilio, P. *Speed and Politics: An Essay on Dromology*, trans. Mark Polizzoni (New York: Semiotext(e), 1986).

Virilio, P. and S. Lotringer, *Pure War*, revised edition (New York: Semiotext(e), 1997).

Walkom, T. 'Replays Show Powell Did Not Score', *Toronto Star*, 11 February 2003.

Washington Times, 'Zarqawi's City of Death', 29 November 2004, p. 20.

Whitehead, Y. 'Information as a Weapon: Reality versus Promises', *Air Power Journal*, (Fall 1997), 40–54.

Wibben, A.T.R. '9.11: Images, Imaging, Imagination', *InfoInterventions*. Available on www.watsoninstitute.org/infopeace/911, 7 October 2001.

5
Risks of Computer-Related Technology

Dr Peter G. Neumann

5.1 Introduction

In this chapter we address computer-related risks associated with, among other topics, individual well-being, world stability, reliability, safety, security, and privacy, and what can be done to combat those risks. In many cases, much greater proactive efforts are needed to reduce the risks. In some cases (as the WOPR computer observes in the movie *War Games*), 'The only winning strategy is not to play'.[1] The topic encompasses risks in defence, aviation, space, control systems, communications systems, finance, health care, information systems generally, and so on.[2]

In light of the long-standing risks of system malfunctions, as well as malicious or accidental system misuse, and impelled by the newly increased awareness of the threats posed by terrorism, it seems natural for us to consider computer-related technologies and their relationships in relation to social, political, economic, and environmental issues. The problems to be deliberated include system security, system reliability, human safety, system survivability, application integrity, privacy, and many other issues.

In that almost everything we do is becoming dependent on computer technologies, for better or for worse, we need to cover a great deal of ground in order to understand what is at stake. The greatest concern here is that we focus on the big picture, without getting lost in less relevant details.

There are numerous dimensions we could consider in what is actually a highly multidimensional problem. Very briefly, some of the dimensional alternatives that come to mind include:

- Internationalism vs isolationism
- Multilateralism vs unilateralism
- Rule by agreement vs rule by force
- Partnerships vs nationalism
- Deregulation vs regulation
- Level economic playing fields vs corporation-dominated globalization

- Free markets vs controlled markets (including international cartels)
- Development of alternative energy sources vs dependence on fossil fuels

However, there are four other sets of alternatives that will be of particular concern to us here:

- Understanding the risks of the misuse of technology vs ignoring them.
- More technology vs less technology for addressing social problems.
- Open information vs secrecy.
- Privacy vs surveillance.

Most of these dimensions are typically considered, rather simplistically, as black-and-white alternatives, sometimes between different ideologies or between 'good' and 'evil'. In reality, things are generally not purely black or white, and we must recognise many shades of grey. Each of these seeming dichotomies is in fact itself a broad range of options, and often simultaneously requires multiple elements along the implied spectrum. Attempts to see everything from one extreme or another are likely to break down, and seem to reflect a serious lack of common sense. Typically, there are no easy answers. I frequently quote Albert Einstein, who said in conversation (although nowhere that I know of in any of his writings) that 'Everything should be made as simple as possible, but no simpler.' As a society, we tend to try to make things too simple, and then belabour our simplistic solutions.

Hence, let us consider how these four dimensions might apply to technology, and in particular to information technology:

5.1.1 Computer-communication technology

For many applications, we need reliable, secure, highly available systems. For many critical applications, we need strength in depth. What we have in practice is weakness in depth. Information systems and networks are riddled with vulnerabilities and weak links. Furthermore, the mass-market marketplace has failed miserably in producing robust systems, although it is wonderful at producing advanced features. We should never assume that the systems on which we depend are invulnerable, or that the people who use and operate them are infallible. We must learn to design systems much more defensively. We must also be suspicious of simplistic folk who say that we can leave it to the marketplace to provide solutions for security and reliability, which are not generally market-force issues.

5.1.2 The Internet

The Internet has opened up enormous new opportunities for global development, worldwide commerce, education, rapid information flow, and so on.

However, the Internet has very little real resistance to coordinated attacks (although what we have seen thus far has more or less been child's play, relative to what could happen), and the systems attached to it tend to be highly vulnerable. Trojan horses, viruses, worms, denial of service attacks, and so on represent real threats, primarily because of the absence of a robust system and network architectures. The beauty of the Internet is that it is truly international. However, the future of the Internet is seriously threatened by its lack of enlightened management, government desires at control, corporate greed, and many other factors. Various Internet taskforces attempt to steer the technological evolution. The Internet Corporation for Assigned Names and Numbers has a fairly narrow charter, but even that has caused enormous controversy. A new organization called People For Internet Responsibility[3] is attempting to encourage more democratic and representative approaches that will ensure that the Internet is truly for everyone, and not ruined by corporate interests, purveyors of electronic junk mail (spam), swindlers, and so on, while at the same time advocating that it is not overly constrained by regulation.

5.1.3 Vulnerability

Our national and international critical infrastructures are riddled with vulnerabilities, including those relating to security, reliability, system survivability, and human safety. This is true of telecommunications, electric power, water supplies, gas and oil distribution, transportation, and even government continuity. For example, the report of the US President's Commission on Critical Infrastructure Protection, under Bill Clinton,[4] concluded that essentially everything is vulnerable to external and internal attacks and indeed to falling apart on its own even without attacks. Many of these risks have been known for many years, although very little has been done substantively in the past.

5.1.4 Openness

There is a big debate within various communities as to whether or not secrecy can increase security. In a few cases, perhaps it can. It is sheer folly to assume that you can avoid the exploitation of serious security flaws by pretending they do not exist. Besides, in the absence of knowledge about how vulnerable you are, you are unlikely to take reasonable remedial measures. Nevertheless, you would prefer that your adversaries do not know more than you do. This is a really nasty problem. The debates over open-source versus closed-source proprietary software are also important. Note that by itself open-source software does not provide the answer. Furthermore, hiding behind flawed proprietary software leads to the institutionalization of security by obscurity, which is inherently a bad idea.

5.1.5 Privacy, secrecy, surveillance, monitoring, oversight: 'Who Watches the Watchers?'

Privacy problems are enormously important, but widely ignored. The average person believes he or she has nothing to hide, so why is privacy important to him or her? Obvious answers include identity theft, false information, monitoring, harassment, blackmail, targeted personal attacks, and many other problems. Furthermore, many privacy problems are institutional rather than individual. In general, some independent oversight is absolutely essential. Companies such as Enron, Global Crossing, Waste Management, and AOL-TimeWarner, to name just a few, have had serious failures of accountability. Anderson's accounting practices have become a poster child for complicity, creative mismanagement, and lack of independent auditing. The situation is much worse with respect to computer systems. Even where independent audit trails exist, they may be tampered with, or destroyed, or bypassed altogether. Although there are often opportunities to reconstruct audit data that has been deleted, there are also serious problems with trying to rely on digital evidence, since the integrity of the evidentiary process may be in doubt. If you have to rely on the integrity of a computer system to protect your information, you are already in trouble, because security problems and privacy violations also involve people who have access to or can penetrate databases. If you have to rely on people who are untrustworthy, all bets are off.

5.1.6 The election process

One example that is not generally recognised as particularly critical involves election processes, which in a sense put many of the previously discussed technological problems such as reliability, security, and privacy into a single context. Many warnings have been given over the past decades, but they have fallen largely on deaf ears. The experience of Florida in 2000 was really just the tip of a huge iceberg. Beginning with the voter registration process, tens of thousands of voters were disenfranchised by bogus felony lists in Florida and other causes – according to the Caltech/MIT study, as many as four to six million votes were lost in 2000. With respect to ballot casting and ballot counting, there are huge risks in the integrity of your vote, the tabulating process, and indeed the accountability of the entire process. Punched cards are clearly problematic. However, all-electronic systems are enormously risky; in today's all-electronic systems, there is no real assurance that the vote you cast is counted correctly, and typically there is no adequate accountability in case of an obvious fraud or unexplained internal error. Such systems offer huge opportunities for fraud. Several of the major companies have been subject to allegations of fraud, including various convictions and serious ethical breaches. The source code is, in almost all cases, proprietary, with the professed nonsensical claim that this makes it more secure. The vendors insist that they are trustworthy and that the systems have been fully tested and certified. However, the testing and certification processes are inherently deficient. Because true democracy depends critically on the integrity of the election process, the old quote is highly relevant: 'It's not who votes that counts, it's who counts the votes.'

The Illustrative Risks document[5] includes numerous pages of cases involving computer-based failures in defence, space, aviation, other transportation, power, medical systems, control systems, the environment, finance, telecommunications, elections, law enforcement, and, perhaps most frustratingly, information security and privacy.

Here are just a few examples from 'Illustrative Risks' and 'www.risks.org', where further details are available.

5.1.7 Commercial aviation problems

- A Lauda Air 767 aircraft had the thrust reversal accidentally deploy in mid-flight.
- During a Northwest Airlines flight, the warning system failed to deploy because it was not powered up.
- A British Midland 737 engine caught on fire, and the pilot shut off the wrong engine (the working engine, rather than the burning one) due to cross-wiring of the actuators.
- An Aeromexico flight crashed into another plane near Los Angeles Airport because of pilot and controller errors.
- Four Airbus A320 crashes were attributed largely to pilot error, although the autopilot and the pilot were not working together.
- An Air New Zealand flight crashed into Mount Erebus in Antarctica because course data was known to be wrong, but had not been fixed.
- A Russian plane was directed to ascend by the automated collision avoidance system, and to descend by the Swiss air-traffic controller, resulting in a crash.

5.1.8 Military and other system problems

- The *Yorktown* missile cruiser was dead in the water for almost three hours as the result of an unchecked divide-by-zero in the application software, which caused the ship's Windows™ operating system to crash. Unfortunately, this had the effect of shutting down the ship's engines.
- The Patriot missile defence system failed to target incoming missiles correctly because excessive drift in the computer clock software placed the incoming objects outside the target area.
- The Aegis system aboard the *USS Vincennes* was unable to distinguish a commercial Airbus from an Iranian fighter plane, and the system's human interface was poorly designed; as a result, the Airbus was mistakenly shot down, resulting in the death of many civilians.

- The Handley-Page Victor aircraft had three supposedly independent tests, all of which seemed to justify the stability of the tail-plane. However, there were three independent failures in those tests: (1) a wind-tunnel model had an error in modelling wing stiffness and flutter; (2) a resonance test was erroneously fitted to aerodynamic equations; (3) low-speed flight tests were incorrectly extrapolated to high-speed flight. The tail-plane broke off in the plane's first flight test, killing the pilot and destroying the aircraft.
- Computerized control systems are increasingly being used in trains, cars, ships, appliances, and so on. Many failures of autonomic controls have been experienced. In addition, numerous train wrecks have been attributed to various combinations of human error, hardware faults, and software problems.

5.1.9 Medical applications

- The Therac 25 (see Nancy Leveson's article and book[6]) radiation machine resulted in several deaths attributed to a critical timing problem in switching from the high-intensity research mode to the low-intensity therapeutic mode.
- A heart-monitor line unwisely fitted with a standard wall plug was mistakenly plugged into the power supply in Seattle Children's Hospital, electrocuting a four-year-old girl in 1986. A similar case occurred seven years later in Chicago.
- Electromagnetic interference has caused accidents and deaths in pacemakers, as have magnets acting on defibrillators.
- Numerous security and privacy problems are associated with medical databases and hospital control systems.

5.1.10 The Year 2000 (Y2K) problem

- A huge lack of foresight led to the Y2K problem, which had been recognised and systematically avoided in systems as early as 1965 (Multics). Enormous resources were expended to prevent serious failures in January 2000. Ironically, some of the supposedly repaired systems failed in January 2001, or later.

More and more systems are *Critical* (for example, *Safety Critical, Survivability Critical*, and so on) as we increasingly computerize and become totally dependent on those systems. Many new risks are being introduced along with new technologies, such as the dependence on biometrics for authentication, and the use of voice-activated speech-understanding systems, subject to native dialects, foreign accents, malicious impersonators, and interference from bystanders. We need to learn more from our own and others' experiences before leaping into would-be technological solutions.

Peter Neumann's website is full of material on how we could dramatically improve the situation. However, a strong personal belief is that no solutions are likely to work in the long run, unless they are based on uncompromising human-oriented democratic principles. Everything we do is becoming inter-related, internationally. This is very obvious when we consider the World Wide Web, the Internet, television, radio, and other media, whereby almost everyone in the civilized world is interconnected one way or another, almost instantaneously.

Perhaps ironically, cartoonists seem to be doing a good job of bringing reality to the public. Take, for example, a quote from George Orwell that appeared in *The Boondocks* comic strip in the Sunday comic pages: 'If liberty means anything at all, it means the right to tell people what they do not want to hear.'[7]

5.2   Roles of technology

We have a tendency to try to solve problems using inappropriate approaches. There are significant risks in attempting to use technological approaches to solve social problems, and similar risks involved in using social, legal or economic solutions to solve technological problems. Beware of uses of technology that appear to offer intopian solutions.

Examples:

- Attempts to prevent terrorism through national missile defence, national identity cards, face scanning, and bombing caves. National identity cards may be seen only as an extension of drivers' licenses, but there are serious risks in the associated databases and infrastructures – identity theft, false arrests, untrustworthy insiders, and so on. Besides, such a card would probably not have prevented the September 11 terrorists, especially those who were masquerading as others, but had what appeared to be legitimate identities. Face scanning generally gives large false positive rates, and at the moment includes only a few dozen faces. Biometrics authentication does have a place in hypercritically sensitive applications, but its general application seems questionable. Typically, we have the problem of putting a safe-like lock on the front door and leaving a side door open. Beware of putting too much faith in these technologies, since many threats can bypass them.
- Attempts to control electronic borders such as telephones, faxes, television, radio, and the Internet: Singapore, China, Taliban, jamming of the Voice of America, and so on.
- Attempts at censorship, for example, by attempting to reject certain types of information such as pornography through simple-minded filtering. Even sillier, the German federal and state governments have recently

agreed to ban pornography world wide, except between 11 p.m. and 6 a.m. in their time zone.

- Attempts to prevent viruses through filters instead of designing systems correctly to prevent them.
- Attempts to control spamming often overzealously block important e-mail.

Technology can do wonders for the entire world, but only if we can move away from simple commercial greed. The commercial marketplace will not solve all of our problems. We cannot dominate and control the world – nor can we be completely isolationist.

We must consider the global implications of everything we do. World economics. The world environment. Combating world poverty and hunger. We pay lip service to better education, but education also seems to be suffering from lowest-common-denominator, increasingly emphasizing the regurgitation of factoids, including misinformation gleaned from the Internet. Creative thinking seems to be deprecated.

Optimizations based on narrow sets of assumptions (what might seem good for me personally? or for my family? or for my company? or for my country?) produce wildly differing results from optimizations based on the realistic assessment of long-range and often not just national implications. Enron is an extreme example of optimizing from the perspective of a few individuals rather than from the perspective of the employees and stock-holders, or even more broadly from the perspective of the good of the nation and the world.

Fossil fuel is another example. Policies based on the assumption that oil is the most important commodity in the world are radically different from human-oriented policies, policies based on alternative energy sources and conservation or moderation. To a man with a hammer, everything looks like a nail. To an investor in oil, everything looks like a dollar bill. To anyone interested in the long-term survival of the planet and the species, conservation looks like a no-brainer.

Long-term research versus short-term profits (as two extreme motivations, with many intermediate approaches). We have become tremendously short-sighted regarding long-term research, which is absolutely essential for the future of the planet. By failing to adequately support essential research, we are eating our own seed corn. There are a few outstanding examples of far-sighted research that has paid off enormously: computer systems, telecom-munications, lasers, biotechnology, and, increasingly, speech recognition and understanding (which are emerging as a huge money saver for the telephone industry). However, in the computer field, and particularly in mass-market software, much of the most important research in robust systems has been ignored in favour of market-driven features. Of course, we are very gifted when it comes to glitsy entertainment and fancy features.

Mass-market software is good at producing dancing pigs on your screen. We produce television sets and other visual media with amazing picture definition, but the content is often lowest common denominator. When it comes to critical systems that must function correctly, securely, reliably, all of the time, the record is amazingly bad.

As a society, we seemed to have evolved into a mentality of 'anything goes as long as you can get away with it'. This seems to be equally applicable to both corporations and individuals, and it has serious effects on the environment and the long-term future of civilization.

We evidently learn little from history. To return to Enron, in January 2002 *The New Yorker* published an article by James Surowiecki on an Enron-like scam from 1861, the Central Pacific Railroad, in which Leland Stanford and his partners set up a contracting subsidiary and scammed the government for at least US$50 million in overcharges. Of course, all of the documents subsequently disappeared.

As a society, we are very bad at reacting in advance to warning signs, although we seem to do fairly well at building new doors after the horse is out of the barn. However, given that our critical infrastructures and our computer-communication technologies are so riddled with vulnerabilities, we need to be much more proactive in future. Unfortunately, the biggest impediment to action seems to be that we have never actually experienced the electronic equivalent of a Pearl Harbor or September 11, and therefore have not been compelled to do enough to protect our infrastructures. This is a characteristic problem for security; unless you have been burned, there is little incentive for proactive action.

We must learn to invest more in the global long-term future, rather than just responding to perceived local short-term needs. Long-term vision is essential. Increasingly, almost everything we do is interrelated; economic policies, energy policies, and technology policies. We should always look at the big picture, rather than just optimizing in the small, and along the way we need much greater altruism.

Open and democratic institutions are clearly our best hope; for nation-states and for technology policy, including ensuring that the Internet evolves constructively. It seems evident that world terrorism is nurtured by almost everything else. However, democracies can be easily corrupted and influenced by intense lobbying. The Enron case might be an example of what could be called 'sweet-and-sour' pork barrels.

With respect to terrorism, a personal favourite mixed metaphor is applicable here: 'we are facing a new era in which Pandora's cat is out of the barn and the genie won't go back in the closet'.

Long ago US Supreme Court Justice Brandeis remarked that governments teach by example. A relevant motto in our actions might be 'Assume that others will do as you do, not as you say.' Hence, we conclude by suggesting that, as individuals and as nations, we need to set consistent examples that have deep commitments to international human rights and human well-being, as well as to economically sound environmental policies. Technology has a significant role to play, if it is used wisely. However, it often further escalates the problems it tries to solve, and sometimes even creates new problems. For example, there is a serious risk of increasing the already huge gap between the 'haves' and the 'have-nots', because technology often benefits only the haves. It also creates a spy-versus-spy spiral where the attackers have a much greater advantage than the defenders. This is certainly true of short-sighted security measures that do not look at the world as a system in the large. Solutions ultimately require pervasive attention to international affairs, rather than just purely domestic considerations.

Notes

1. L. Lasker and W.F. Parkes, *War Games*, USA, 1983.
2. For extensive background information, see the following: Peter Neumann website: http://www.csl.sri.com/neumann; The Illustrative Risks compendium indexes to Risks cases: http://www.csl.sri.com/neumann/illustrative.html; and the archives of the Risks Forum: http://www.risks.org.
3. People for Internet Responsibility: http://pfir.org/.
4. Department of Justice, *White Paper: The Clinton Administration's Policy on Critical Infrastructure Protection: Presidential Decision Directive 63*, 22 May 1998. Available at http://www.usdoj.gov/criminal/cybercrime/white_pr.htm.
5. Peter Neumann website: http://www.csl.sri.com/neumann/illustrative.html.
6. For example: N.G. Leveson, and C.S. Turner, 'An Investigation of the Therac-25 Accidents', *IEEE Computer*, 26 (7) (July 1993), 18–41.
7. A. McGruder, 'The Boondocks', *San Francisco Chronicle*, 30 January 2002.

6

Missile Defence – The First Steps Towards War in Space?

Professor David Webb

6.1 The military use of space

The term 'Revolution in Military Affairs' (RMA) represents a move to bring together technologies to aid war fighting and management. Over the last few decades the US military in particular has developed technological systems to enable as near a perfect condition of 'situational awareness' as is possible in any arena of conflict. These command-and-control systems and computer networks have been integrated through the use of space-based technology.

This chapter will argue that the USA sees the RMA reliance on space as requiring an active space defence system. Such a system is outlawed in spirit and deed by government statements and international treaties, and so the events of September 11 and the 'war on terror' have supplied a useful opportunity to develop defensive space warfare systems (under the guise of missile defence) as a step towards the implementation of offensive systems to attack others should the need arise.

The use of space by the world's military is now well established and has become indispensable for the USA which has built on a range of experiences over a period of years in:

- Operation Desert Storm, Kuwait, Iraq, 1991.
- Operation Allied Force, Kosovo, 1999.
- Operation Enduring Freedom, Afghanistan, 2002.
- Operation Iraqi Freedom, Iraq, 2003.

During Operation Iraqi Freedom US satellite information allowed a military response in minutes rather than the hours or days it had taken previously. This shortening of the so-called '*kill chain*' means that space has now become '*the ultimate military high ground.*'

In a typical battle situation the US military now relies on space-based weather prediction systems (the Defence Meteorological Support Program), military communications satellites (MILSTAR – to communicate from command centres and between troops), espionage and surveillance satellites (to intercept communications by an adversary and collect images of troop movements and weapon placements), early warning satellites (to provide information on missile launches) and military Global Positioning System (GPS) satellites to allow troops and vehicles to navigate and to quickly and accurately specify targets and guide '*smart*' bombs and unmanned aerial vehicles (UAVs).

The USA deployed 6,600 GPS-guided munitions and over 100,000 Precision Lightweight GPS Receivers in the Iraq War.[1] The US military was using ten times the satellite capacity that it had used in the Gulf War of 1991. Nine days before the start of the war a new US Defense Satellite Communications System was installed to integrate US military forces on land, sea and air with the Pentagon, the White House, the State Department and the US Space Command. Over 100 military satellites supported the US and UK war effort, 27 GPS satellites were available to help determine the exact location of special operations teams and of targets, and around 24 communications satellites offered command and control and gave warnings of missile attacks. There were also weather forecasting, TV and other systems in operation. A February 2000 flight of the space shuttle *Endeavor* used radar to produce a three-dimensional map of targets in Iraq.[2] The human resources available are also extensive; Director of Space Operations Major General Judd Blaisdel estimated that 33,600 people at 36 sites around the world were involved in space-war activities.[3]

This massive increase in the use of space technology for war fighting is not without problems. Modern weapons systems also require that communications satellites carry enormous amounts of traffic. For example, one Global Hawk UAV requires about 5100 megabits per second of bandwidth (five times the entire bandwidth required by all of the US military during Desert Storm) and in fact in 2002 the *Wall Street Journal*[4] reported that during 'Operation Enduring Freedom' the Pentagon could only deploy four (one half) of its available UAVs at any one time because there was not enough bandwidth available to fly all of them. Future requirements are likely to be overwhelming; a Defense Science Board study has predicted that by 2010 the Pentagon will require 16 gigabits per second of bandwidth to support a major war.[5]

Of course, the USA is not alone in its use of space for military purposes. Russia has a number of military satellite programmes, with five types of short-lifetime imaging reconnaissance satellite which can be launched to update topographic and mapping data and two series of electronic intelligence (ELINT) satellites. It also has four types of dedicated military communications satellites, with around 24 being launched since 1997 (although some of these are no longer functioning).[6] Russia also has a number of navigation satellites and a dual-use Global National Satellite System (Glonass) which is similar to GPS.[7] The Russian armed forces are also to be outfitted with Glonass receivers by 2005.[8] There are, in addition, Russian ballistic missile early warning and space monitoring systems.

The military use of space is proliferating rapidly. China has launched a number of military satellites, India has imaging and communication satellites suitable for military use and Israel has military satellites, has plans for new communications, imaging and radar satellites and is considering a system that would allow launch on demand of small satellites from fighter aircraft.[9] Other countries, such as Brazil, Pakistan and Ukraine, have military space capability or potential,[10] Australia has a dual-use military-commercial communications satellite[11] while in Europe the UK, France and Italy make extensive use of military satellites for imaging and communications and the European Space Agency (ESA),[12] set up to be an entirely independent organization, is slowly becoming politicized (with increasing control being exercised by the European Commission)[13] and even possibly militarized through its Galileo GPS system.[14]

6.2   Anti-satellite (ASAT) programmes

This reliance on space for command, control, communications, computer, intelligence, surveillance and reconnaissance (C4ISR) has one serious disadvantage: space-based satellite systems are extremely vulnerable to attack from anti-satellite (ASAT) systems. Shortly before his appointment as Secretary of Defense, Donald Rumsfeld chaired the 'Commission to Assess United States National Security Space Management and Organization'[15] which concluded, in January 2001, that the likelihood of an attack on US space systems needed to be taken seriously to prevent a future 'space Pearl Harbor'.

In fact, the first actual attack on any military satellite system occurred in 2003 when the Iraqi military unsuccessfully attempted to jam the US GPS.[16] US Air Force Secretary James Roche commented that this attempt to disrupt GPS-guided weapons demonstrated the world's understanding of the importance of space to the US military. Interestingly, in 2004 the US Air Force itself deployed a number of reversible jamming, or Counter Communications, systems.[17] However, a more threatening scenario is the possibility of the deployment of actual weapons systems against satellites.

Since the beginning of the space age, Russia and the USA have both openly worked on several ASAT projects. Initial efforts in the 1950s consisted of well-known air-launched missile technology, but more sophisticated systems have since been developed.

6.2.1   The Soviet Union and Russia

In the 1960s the Soviet Union surrounded Moscow with nuclear-tipped intercontinental ballistic missiles to act as an Anti-Ballistic Missile (ABM) system. These missiles would also have ASAT capabilities as they would be able to destroy all space-based systems in the vicinity of their detonation. However, the main ASAT system developed by the Soviet Union was the 'Co-orbital ASAT' – a kamikaze satellite packed with explosives. Development of the *Istrebitel Sputnikov* (fighter satellites) began in the early 1960s and the first test flights were made in 1968. The ASAT was to be placed in an orbit close to that of the target and would move in to destroy it within one or two orbits. Initial tests made between 1963 and 1972 indicated that the system could work from altitudes of 230 to 1,000 kilometres and the system was declared operational.

The Soviets temporarily ceased testing the system after signing the ABM Treaty in 1972, but they resumed again in 1976 and continued until 1982. During this time the effective range of the system was reportedly extended to altitudes of 160 to 1,600 kilometres.[18] In 1983, the Soviet Union declared a moratorium on launching ASATs, on condition that no other country deployed such a system, and Russia seems to have continued to observe this policy.[19] Jane's 2001–2 *Space Directory* describes the Russian ASAT programme as '*inactive*'.

6.2.2 USA

The USA began tests in 1959, but results were not encouraging and the project was stopped in 1963, although related US Navy projects did continue into the early 1970s. In the 1960s the destruction of satellites by the use of nuclear explosions was considered. A 1.4 megaton high-altitude nuclear test explosion detonated 400 kilometres over the Pacific in 1958 did damage three satellites. However, the potential damage to untargeted areas and systems through radiation and the electromagnetic pulse (EMP) meant that no actual ASAT tests of this type were carried out although the nuclear-carrying *Nike Zeus* was adapted for ASAT use from 1962. A single-missile ASAT was deployed at Kwajalein Atoll in the Pacific until 1966 under Program 505 codenamed 'Mudflap' and was then replaced by the USAF Thor ASAT until 1972.[20]

The resumption of USSR ASAT tests in 1976 could have come about as the result of reports of a renewed US interest in ASAT technology and the development of the US Space Shuttle programme (which was considered to have an ASAT capability). The US was itself concerned with exaggerated reports of Soviet laser and particle beam ASAT/ABM technology and revived its ASAT programme with the Air-Launched Miniature Vehicle (ALMV) which was fired from an F-15 aircraft, carried a heat-seeking homing device and was designed to attack Low Earth Orbit (LEO) satellites. The missile consisted of a modified Short Range Attack Missile (SRAM) first stage, a Thiokol Altair III second stage, and a Vought miniature homing vehicle (MHV). The ASAT was launched at high altitude from an F-15 aircraft in a steep climb. This gave the ASAT rocket a useful initial velocity to allow it to reach its target in orbit. After the first stage separated, the second stage would propel the MHV into space on a collision course with the target satellite which was destroyed kinetically by ramming at high speed. The US carried out five tests from 1984 to 1986 and actually tested the system against a satellite in September 1985.[21] However, considerable cost increases for further development led to the programme being cancelled in 1988. In the same year, the US Congress voted against extending a unilateral ban on ASATs and development started on new ASAT systems.

Under President Reagan's 1983 Strategic Defense Initiative (SDI) ASAT projects were adapted for ABM use and vice versa. Initially, the plan was to use the MHV as a basis for a collection of about 40 space platforms containing up to 1,500 kinetic interceptors. By 1988 the project had evolved into an extended four-stage development. The first stage was the *'brilliant pebbles'* system consisting of many single kinetic interceptors and associated tracking systems. The second stage would deploy larger platforms and the following phases were to include laser weapons and, later, charged particle beam weapons. Plans were to complete the whole thing by 2000 at a cost of around $125 billion.

The only successful energy weapon to emerge from the SDI project was the Mid-Infrared Advanced Chemical Laser (MIRACL).[22] This can produce a megawatt of output for around 70 seconds and was developed mainly in response to intelligence that the Soviet Union had created a similar system. However, after an official US visit to the Soviet Union in 1989 discovered that the Soviet system was no threat and that it was far from completion, Congress banned the use of MIRACL in 1991. The development of the US Army ground-based kinetic energy ASAT (KE-ASAT) system was also banned in 1993, but this was resurrected in 1996 with US$45 million of funding which continued until 2002.

In 1996 the ban on using the MIRACL ended and the following year the system was tested by firing at a USAF satellite 420 kilometres above the Earth – supposedly to see if US satellites could withstand a laser attack. Currently, the KE-ASAT needs more funding and testing before it can become operational. The ALMV has not been tested and there appears to be little interest at present in reviving the system. The MIRACL laser is being further developed with Israel, but it has not been tested since 1997 and its full capabilities are not known.

6.2.3   China

At the time of writing, China does not have a publicly declared ASAT programme, although their existing launch capabilities could provide the basis for the development of such a system.[23] A programme to field a viable ASAT system consisting of a kinetic kill vehicle, high-powered laser, space early warning, and target discrimination system components, was abandoned in 1980. Preliminary research on ASATs has been carried out since then, partly funded under a Program for High Technology Development.[24]

In 2003 and 2004, the annual reports to the US Congress on Chinese Military Power quoted an article from a Hong Kong newspaper that reported China as having developed and tested a 'parasitic micro satellite' ASAT system. However, this information seems to have originated from an item posted in 2000 on an unreliable Internet bulletin board service run by a self-described 'military enthusiast'.

6.3 Current US developments

The USA has recently shown an increase in funding and support for ASAT and related programmes. In 2004 the Pentagon received US$168.6 million for the development of space weapons technology and over US$2 billion for weapons-related programmes.[25] In August 2004, the USAF released a document, entitled *Counterspace Operations, Air Force Doctrine Document 2–2.1*,[26] which details, for the first time, US anti-satellite and space weapons operations. The Foreword by General John P. Jumper, USAF Chief of Staff, states that 'U.S. Air Force counterspace operations are the ways and means by which the Air Force achieves and maintains space superiority. Space superiority provides freedom to attack as well as freedom from attack... Space and air superiority are crucial first steps in any military operation.' The document discusses air-launched missiles, direct-ascent and on-orbit ASATs as possible mechanisms for destroying satellites.

The Pentagon budget request for space control and space force projection-related programmes for 2005 totalled over US$3 billion, including around $217 million for potential ASAT and space-weapons-associated projects. The appropriations committees cut nearly US$1 billion from the military space budget[27] and sliced 40 per cent from the space weapons and ASAT requests. The agreed budget includes US$10.6 million for initial work on the space-based interceptor test bed,[28] but the Congressional appropriators directed the Force Application and Launch from the Continental US (FALCON) programme[29] not to engage in any 'weapons-related work' during fiscal year 2005 and cut funding for the Common Aero Vehicle (CAV) by a half to $12.5 million (any effort to put weapons on the CAV or test launch it on a ballistic missile was also forbidden).Other space programmes suffered funding cuts from appropriators, including the Space-Based Radar (SBR), Transformational SATCOM (T-SAT) and Counter Surveillance Reconnaissance System (CSRS) programs.

A recent article in *Aerospace Daily & Defense Report*[30] quotes a scientist at Science Applications International Corporation (SAIC) as saying that these cuts are: 'largely due to the concern over the proper use of force in space and the vocal anti-space weapons community.' 'To their credit, they have been on the field. The people who are advocates of the funding for these particular programs...haven't well engaged in that debate.' Peter Huessy of the National Defense University Foundation was also quoted as saying that the anti-space weapons lobby has been effective in part because of its significant financial backing. The lobby is 'being led, unfortunately, by not just the traditional arms control community, but about $100 million a year from foundations,' according to Huessy. 'And that kind of money is so far and beyond anything being spent by the proponents.'

Of course, this is probably just an argument for more funding for lobbying and no doubt that will happen, but it appears that the activities of nongovernmental organizations (NGOs) such as the Center for Defense Information (CDI),[31] the Center for Nonproliferation Studies and the Monterey Institute of International Studies (CNS/MIIS),[32] assisted by grass-roots campaigners such as the International Network of Engineers and Scientists Against Proliferation (INESAP), the Global Network Against Weapons and Nuclear Power in Space[33] and the Campaign for Nuclear Disarmament,[34] may have been effective.

Another project to come up against a funding hitch is the controversial Near Field InfraRed Experiment (NFIRE) of the Missile Defense Agency (MDA), whose primary role is to gather data to help differentiate between the rocket and its exhaust plume. The proposal was to launch a platform termed a 'kill vehicle' to closely encounter a target missile with an obvious capability to disable or destroy targeted missiles or orbiting satellites. The NFIRE was originally to be launched from a Minotaur missile in summer 2004, but the MDA announced in March of that year that there would be a year-long delay apparently due to having received only US$44.5 million of the requested US$82 million of funding in 2004. Then, in July 2004, the Congressional appropriators cut the entire US$68 million budget requested for NFIRE, although the Senate Appropriations Committee recommended that the programme should be preserved. It is now scheduled to be launched in late 2005 or early in 2006[35] and it was reported in *Space News* in August 2004 that the controversial sensor (the 'kill vehicle') would be removed from the programme. The report stated that: 'U.S. Rep. Loretta Sanchez, D-Calif, championed the effort to persuade Pentagon officials to consider restructuring the NFIRE program to exclude the kill vehicle. "My biggest concern," Sanchez said last week, "was what message we might send to other nations."'[36]

One other growth area with a clear ASAT capability is the ongoing development and testing of US Micro-Satellite (MS) prototypes, including a 28-kilogram XSS-10 MS to manoeuvre around and photograph space objects.[37] The USAF launched the first satellite of these in January 2003. A larger version, the XSS-11, will remain in orbit for one year and will transmit real-time streaming video to ground stations. The 'single strongest recommendation' of the informal Air Force 1999 MS Technology and Requirements Study, was for 'the deployment, as rapidly as possible, of XSS-10-based satellites to intercept, image, and if needed, take action against, a target satellite'.[38]

Much of the current US development of space-based technology and weaponry (including space-based interceptors)[39] and air-borne and space-based lasers) is taking place under the missile defence umbrella. As David Wright and Laura George from the Union of Concerned Scientists have stated:

> ...current US ASAT capability is fairly limited and, based on current funding levels, dedicated ASAT systems appear not to be high priorities. Some of the planned missile defence systems, on the other hand, would add significant ASAT capability to the US arsenal and have strong political and financial support. This fact should be kept in mind when analysing US capabilities and developing policies relevant to restricting ASATs.[40]

It seems clear then that projects that are overtly developing ASAT or space weapons systems are having some difficulties obtaining funds. However, there are other ways of obtaining large sums of money for very similar space-based programmes under the guise of a system to defend against missile attack from terrorists or 'rogue states'.

6.4 Missile defence

US missile defence is justified to the American people and the world as a defence against limited missile attack, but it can also be seen as a proving ground for certain space weapon components. In the past any strategy for ballistic missile defence (as opposed to theatre missile defence which involves short-range battlefield weapons) has been formulated in the light of traditional nuclear deterrence theory between major nuclear states and the concept of 'mutually assured destruction'. Currently, however, the USA is justifying development of ballistic missile defence systems by pointing to a threat from 'rogue' states such as North Korea and Iran, or from terrorist groups. However, there is no evidence that any of the states specified has the technology to launch a long-range missile at the USA (or even the intent – such an act would be suicidal) and there are numerous alternative means (that are cheaper and easier to obtain) for terrorist groups to deliver a nuclear weapon or dirty bomb.

There is an alternative view as to why the US is so keen to develop technologies that are unreliable, costly and controversial. While it may be difficult to convince people that space weapons are necessary, it may be much easier to generate a fear of attack from terrorists or 'rogue' states in order to justify the development of space weapons technologies. There are a number of examples of technological systems being developed for missile defence that can easily be adapted to war fighting or ASAT roles.

The development of improved space-tracking facilities on the ground (such as the upgrading of the Ballistic Missile Early Warning System (BMEWS), radars at Fylingdales in the UK and Thule in Greenland and the development of the X-band radar) and in space (such as the SBR, and Space Tracking and Surveillance Systems (STSS), or Space Based Infra Red System (SBIRS)) is integral to missile defence, but could also be used for ASAT capabilities. Interceptor missiles for the ground-based mid-course defence element of Missile Defence, designed to hit and intercept incoming missiles, could also be deployed against LEO satellites.[41]

The Air-Borne Laser (ABL) – a high-powered laser fitted to a modified Boeing 747 which is under development and being tested – is capable of both intercepting missiles and destroying, or at least blinding, satellites.[42] Although the Space-Based Laser (SBL) programme has been more or less cancelled, the idea of a powerful land-based laser using a space-based mirror system has been proposed to act as a missile defence and/or space weapon.

There are also a number of reasons why the pursuit of a national missile defence programme can be questioned. Firstly, of course, why would any group or nation use an expensive and unreliable long-range missile to deliver a nuclear or similar weapon to the US when it is cheaper, easier and more trustworthy to smuggle a device into the US by way of a boat or truck?

Also, as long ago as 2000, the Union of Concerned Scientists[43] made it clear that any group capable of launching an Intercontinental Ballistic Missile (ICBM) with a nuclear warhead would also be capable of deploying sufficient countermeasures to overcome the missile defence system as proposed by the Pentagon. The question also remains as why a state or terrorist group would launch an expensive and unreliable long-range missile rather than deliver such a weapon by stealth – by truck or in a ship?

Another scientific study by the American Physical Society (APS)[44] concentrated on boost-phase interception. This form of ABM system is favoured by many US military strategists for a number of reasons; for example, it would enable interception before the deployment of decoys and would mean that debris would be scattered over the launch area rather than the target area. The APS study showed that ground- or air-based interceptors (even ABLs) would need to be positioned too close to the launch sites to be practical and that a huge fleet of around 1,600 space-based interceptors would be needed to guarantee global coverage at all times. Even if an intercept was able to be made in the first few minutes of launch (during the burn time of the missile's rocket motors), it is still likely that a missile launched from somewhere in the Middle East, say, would reach the American continent.

In December 2002 President Bush ordered the deployment of the first ten long-range interceptors by the end of 2004. Six interceptors are now installed at Fort Greely in Alaska and four more are being placed at Vandenberg Air Force Base in California. These are to be followed by ten more in each place in 2005, with the possibility of hundreds more in other sites (some possibly outside the USA) before 2012. The Pentagon originally set the date of 30 September 2004 for the activation of the system, but there have been numerous hold-ups and continual worries about an over-hasty implementation.

With hundreds of billions of dollars already spent on developing the associated system components, there are also growing concerns about the overall costs. It has been estimated that the research and development (R&D) costs for the period 2004–9 will be approximately US$50 billion, with the total lifecycle cost for a layered missile defence system close to US$1.2 trillion up to 2035.[45] However, the USA is doing its best to involve as many others as possible, as is outlined in the following section.

6.4.1 International response to US missile defence

In Europe, three countries – the UK, Denmark and Greenland – have already agreed to become part of the missile defence system by allowing the early warning and tracking radars at Fylingdales and Thule to be upgraded for this purpose. In July 2003 a Missile Defence Centre was launched in the UK and two months later it was revealed that the British government is to spend £5 million a year until 2009 on missile defence. Other European countries are being tempted with associated R&D contracts. For example, a US$3 billion contract has been awarded to the MEADS (Medium Extended Air Defense System) missile defence system – a joint venture between Lockheed Martin Corp, the European Aeronautic Defence and Space Company (EADS), MBDA-Italia and Lenkflugkorpersysteme (LFK), Germany. The USA has also been negotiating with Poland and the Czech Republic over the possibility of sitting radar stations and/or missile interceptor sites in those countries. Hungary, Romania and Bulgaria have also been subject to similar approaches.

On 7 July 2004 Australia signed a framework memorandum of understanding with the USA, outlining future Australian participation on cooperative missile defence development and testing over the next 25 years. The agreement aims to establish new joint efforts and includes specific arrangements for collaboration on the development and testing of advanced radar technology for the improved early detection of ballistic missiles and the potential options for a missile defence capability for a new Australian destroyer.

At the time of writing, Canada is alone in pulling away from involvement in US missile defence, although Ottawa has already agreed to amend its NORAD agreement with the USA so that its missile warning function is available to the US missile defence system.

Japan was the first country to agree to work with the USA on ship-based missile defence because of the perceived threat from North Korea. In 1998 Japan was shocked by the launching of a North Korean multi-stage 'Taepodong' ballistic missile over Japan's main island and, although in 2002

Japanese Prime Minister Junichiro Koizumi won a promise from North Korea for a moratorium on further long-range testing, distrust runs deep and Tokyo is responding by upgrading its own destroyers and acquiring US-made interceptors.

Russia has stated that it would be prepared to cooperate with the US as long as an agreement on the demilitarization of space could be reached. However, this is unlikely and so Russia is continuing to go it alone with the further development of its missile defence systems and countermeasures. In February 2004 President Putin announced that Russia had successfully tested a new nuclear hypersonic missile that is capable of altering course as it nears its target. This was a clear statement that new types of missile were being developed to overcome missile defence systems, justifying the criticism that missile defence would lead to another technological arms race.

China has expressed concern over the implementation of missile defence, being particularly concerned about the potential deployment of land- or ship-based missile defence systems in the Pacific region by Taiwan, South Korea and Japan.

The widespread development of missile defence systems (including Theatre Missile Defence) and the possible use of such systems for fighting wars in and from space have led to the popular terminology of 'Star Wars' to generically describe these activities.

6.5   The possibility of space weapons control

6.5.1   What is a space weapon?

Given the US enthusiasm for missile defence and its attempts to justify it worldwide, despite the possible escalation in space weapons development, ways of preventing a war in space must be found. One major problem here though is the definition of a space weapon. There are many space-based systems that could be employed for useful nonmilitary purposes or just as easily turned to an offensive capability. For example: manoeuvrable micro satellites or space planes. There are also space-based components such as communications or surveillance systems (or mirrors to direct ground-based lasers) that, although not weapons themselves, could be part of a weapons targeting or battle management system.

Neuneck and Rothkirch have pointed out that definitions of space weapons could be technical, geographical or politically motivated.[46] At the Conference for Disarmament In 1985 China proposed that space weapons be defined as 'all devices or installations based in space (including those based on the Moon and other celestial bodies) which are designed to attack or damage objects in the atmosphere or on land, or at sea'.

It is surely not beyond the wit of humankind to come to some agreeable definition?

6.5.2 Treaties

The 1967 Outer Space Treaty[47] provides the basic framework for international space law. Among other things, it recognises that:

- the exploration and use of outer space shall be carried out for the benefit and in the interests of all countries and shall be the province of all mankind;
- outer space shall be free for exploration and use by all states;
- outer space is not subject to national appropriation by claim of sovereignty, by means of use or occupation, or by any other means;
- states shall not place nuclear weapons or other weapons of mass destruction in orbit or on celestial bodies or station them in outer space in any other manner;
- the Moon and other celestial bodies shall be used exclusively for peaceful purposes.

Other relevant space treaties include the Agreement Governing the Activities of States on the Moon and Other Celestial Bodies (the Moon Agreement) which entered into force in 1984. The 1932 International Telecommunication Union (ITU) Convention, amended in 1992 and 1994, protects civilian satellites from interference.

So it is possible to come to international agreement on many issues concerning the use of space. However, there is no current treaty that prevents the stationing of weapons in space – other than weapons of mass destruction – and a treaty has not yet been negotiated to comprehensively prevent an arms race in outer space.

In the United Nations (UN), there is general agreement, including among all space capable countries, that an arms race in outer space should be prevented. The Committee on the Peaceful Uses of Outer Space (COPUOS) attached to the General Assembly's Fourth Committee (Special Political and Decolonisation), and UNISPACE holds periodic meetings, but usually discusses space exploration issues. Military-related problems in space are discussed by the First Committee (Disarmament and International Security) and negotiations on these issues are held within the Conference on Disarmament (CD).

Since 2002 two major space arms control initiatives have been discussed within the CD; a treaty to ban space weapons (led by China and Russia) and the creation of a Prevention of an Arms Race in Outer Space (PAROS) ad hoc committee. In the past China has insisted that any such committee must include the creation of a space weapons ban treaty as its mandate and has also tied its agreement on negotiations for a Fissile-Material Cutoff Treaty to the negotiation of a PAROS treaty. On the other hand, the USA is the key opponent to a treaty banning space weapons, but does favour the creation of a PAROS ad hoc committee, so long as the committee's mandate is broad.

Although there is general agreement on the need for a PAROS ad hoc committee, differences over its proposed mandate have continuously stalled consensus within the CD.

In 2000 the UN General Assembly adopted a PAROS resolution by a vote of 163 to none, with three abstentions. The three states that abstained were the Federated States of Micronesia, Israel and the United States of America.[48] More recently, on 20 October 2004, at the UN First Committee (Disarmament and International Security) in New York, a number of states highlighted the importance of preventing the deployment of weapons in outer space. The previous day, at a special session, Russia had announced its new policy of 'no first deployment' of space weapons and it had also joined with China to submit a draft treaty to prevent the placing of weapons in space. Egypt and Sri Lanka also introduced their traditional PAROS resolution emphasizing 'the need for further measures with appropriate and effective provisions for verification to prevent an arms race in outer space' and calls on the CD to establish a PAROS ad hoc committee in its 2005 session. The vote was 167 for, none against and two abstentions (the US and Israel).[49]

Among all these discussions and debates, politics and military or industrial advantage are always at the forefront. Hardly ever does an ethical argument enter into these discussions, although it could be said that perhaps the most successful space treaty of all, the Outer Space Treaty, declaring space as the province of all human kind, is basically ethical. Perhaps it is time to turn to an ethical consideration of the problem in order to find a common human resonance and a path through vested interest, fear and mistrust?

The ethical use of space was considered at a conference entitled 'Space Use and Ethics. Criteria for the Assessment of Future Space', held in 1999 at the Darmstadt University of Technology (TUD) in Germany. In the twenty-first century, space technology should contribute to solving conflicts and problems on Earth in a sustainable way.[50]

In order to assess the use of space technology and to ensure its societal acceptance, costs and resources, goals and benefits, and also undesired consequences and risks, Jürgen Scheffran from INESAP suggested eight concrete criteria for the assessment of future space projects which can also be applied to other fields of technology:

- Exclude the possibility of severe catastrophe.
- Avoid military use, violent conflict, and proliferation.
- Minimize adverse effects on health and environment.
- Assure scientific-technical quality, functionality, reliability.
- Solve problems and satisfy needs in a sustainable and timely manner.
- Seek alternatives with best cost–benefit effectiveness.
- Guarantee social compatibility and strengthen cooperation.
- Justify projects in a public debate involving those concerned.

At a time when satellite- and missile-related technologies seem destined to grow rapidly in a relatively unconstrained manner, the nations of the world need to take these issues very seriously and come to some agreement on how to move forward in a way that can best guarantee human survival. Missile defence leading to more missile offence and space weapons cannot be the path to follow.

Notes

1. Jeffrey Lewis, 'What if Space Were Weaponized?,' Center for Defense Information, Washington DC.
2. As reported in the *Colorado Springs Gazette*, 13 April 2003.
3. Michael Woods, 'Satellites Provide Vital Reconnaissance, Communications to War Effort,' Post-Gazette. Available at, http://www.post-gazette.com/nation/20030402spacewar0402p4.asp.
4. Greg Jaffe, 'Military Feels Bandwidth Squeeze As the Satellite Industry Splutters,' *Wall Street Journal*, October 4 2002.
5. As reported in Lewis, 'What if Space Were Weaponized?'
6. See 'Current and Future Space Security – Russia: Military Programs' from the Center for Nonproliferation Studies, Monterey Institute of International Studies – http://cns.miis.edu/research/space/russia/mil.htm.
7. www.fas.org/spp/guide/russia/nav/glonass.htm.
8. Nikolay Poroskov, 'Platoon With a Satellite', *Vremya Novostey*, 21 August 2003; in 'Russian General Staff Approves Plan to Equip Troops With GLONASS Navigation Receivers', quoted in http://cns.miis.edu/research/space/russia/mil.htm.
9. Barbara Opall Rome, 'Israel Makes Plans for Broad Space Capabilities,' *Space News*, 25 August 2003.
10. 'Countries with Advanced-Launch Capabilities' from the Monterey Institute of International Studies – see: http://cns.miis.edu/research/space/spfrnat.htm.
11. 'France Launches Australian MilSat Half Owned by Singtel', spacedaily.com, 11 June 2003, http://www.spacedaily.com/news/milspace-comms-03t.html.
12. See European Space Agency, www.esa.int, in particular, www.esa.int/esaCP/SEMFEPYV1SD_index_0.html.
13. See, for example, Regina Hagen, 'Europe – the Leading Space Power?,' INESAP Bulletin no. 23, April 2004, http://www.inesap.org/bulletin23/art04.htm.
14. 'An Evaluation of the Military Benefits of the Galileo System' by James Hasik and Michael Rip, GPS World, April 2003. Available at http://www.gpsworld.com/gpsworld/article/articleDetail.jsp?id=53279.
15. Available from: http://www.defenselink.mil/pubs/space20010111.html.
16. 'Jamming Incident Underscores Lessons About Space', spacedaily.com. Available at http://www.spacedaily.com/news/gps-04zzzzb.html.
17. Jim Wolf, 'US Deploys Satellite Jamming System,' Reuters, SanDiego.com, 29 October 2004.
18. Laura Grego, Union of Concerned Scientists, 'A History of US and Soviet ASAT Programs,' April 2003. Available at http://www.ucsusa.org/global_security/space_weapons/a-history-of-asat-programs.html.

19. Aleksandr Dolinin, Interview with Space Troops Commander Colonel-General Anatoliy Perminov, 'Outer Space and the Military Security of Russia', *Krasnaya Zvezda*, 27 April 2001, p. 1; in 'New Space Troops commander Colonel-General Anatoliy Perminov interviewed on connection between Space Troops' activities and various areas of country's development.'

20. See http://www.paineless.id.au/missiles/NikeZeus.html and http://en.wikipedia.org/wiki/Anti-satellite_weapon.

21. The USA tested the ALMV against an ageing Solwind satellite in a 555km orbit on 13 September 1985.

22. For more information, see: http://www.fas.org/spp/military/program/asat/miracl.htm.

23. 'Chinese Anti-Satellite Capabilities' from GlobalSecutiy.com at http://www.globalsecurity.org/space/world/china/asat.htm.

24. Ibid.

25. Jeffrey Lewis, 'Space Weapons Spending in the FY2005 Defense Budget,' presented at the 9th PIIC Beijing Seminar on International Security, October 2004.

26. Available from: http://www.dtic.mil/doctrine/jel/service_pubs/afdd2_2_1.pdf.

27. Ibid.

28. Jeffrey Lewis, 'Programs to Watch', in 'Weapons in Space', Arms Control Today, November 2004, available at: http://www.armscontrol.org/act/2004_11/Krepon.asp.

29. Details of the FALCON programme can be found in the 'US Air Force Transformation Flight Plan', November 2003 – see http://www.af.mil/library/posture/AF_TRANS_FLIGHT_PLAN-2003.pdf.

30. Jefferson Morris, 'Space Weapon Proponents Need to make Better Case', *Aviation Week's Aerospace Daily & Defense Report*, 211 (25).

31. http://www.cdi.org.

32. www.cns.miis.edu.

33. www.space4peace.org.

34. www.cnduk.org.

35. 'US Might Intercept Target from Space in 2006' from the Global Security Newswire. Available at http://www.nti.org/d_newswire/issues/2004_4_29.html#DAE4AB71.

36. Jeremy Singer, *Critics Laud Plan to Remove 'Kill Vehicle' From Satellite*, space.com. Available at http://www.space.com/spacenews/archive04/nfirearch_082304.html.

37. Theresa Hitchens and Jeffrey Lewis, 'Arms Race in Space?' US Air/force Quietly Focuses on Space Control', *Defense News*, 1 September 2003.

38. Matt Bille, Robyn Kane, and Maj. Mel Nowlin, 'Military Microsatellites: Matching Requirements and Technology', presented to the AAIA Space 2000 Conference and Exhibition, Long Beach, CA, 19–21 September 2000, p. 9.

39. The US has stated its intent to launch a space-based interceptor test bed by 2008. See, for example, Theresa Hitchens and Victoria Samson, 'Space-Based Interceptors – Still Not a Good Idea', Center for Defense Information, Summer/Fall 2004 – available at www.cdi.org/news/space-security/space-based-interceptors.pdf

40. David Wright and Laura George, 'Anti-satellite Capabilities of Planned US Missile Defence Systems', *Disarmament Diplomacy*, Issue No. 68, December 2002–January 2003. Also at http://www.acronym.org.uk/dd/dd68/68op02.htm and from the Union of concerned scientists: http://www.ucsusa.org/global_security/space_weapons/page.cfm?pageID=1152.

41. Ibid.

42. Ibid.

43. See: http://www.ucsusa.org/global_security/missile_defense/index.cfm.
44. 'Boost-Phase Intercept Systems for NMD', a report by the American Physical Society Study Group, July 2003.
45. R.F. Kaufman (ed.), 'The Full Costs of Ballistic Missile Defense' Center for Arms Control and Non-Proliferation, January 2003.
46. For a more detailed discussion see G. Neuneck and A. Rothkirch, 'Space as a New Medium of Warfare? Motivations, Technology and Consequences', Institute for Peace Research and Security Policy, University of Hamburg.
47. See: http://www.oosa.unvienna.org/SpaceLaw/outerspt.html.
48. More details in the UN Press Release GA/9829 – available at http://www.un.org/News/Press/docs/2000/20001120.ga9829.doc.html.
49. Rebecca Johnson, 'PAROS discussions at the 2004 UN First Committee', The Acronym Institute, 20 October 2004. Available at http://www.acronym.org.uk/un/2004paro.htm.
50. See reports from the conference in INESAP bulletin no. 17. Available at http://www.inesap.org/bulletin17/bul17art19.htm.

7
Technology as a Source of Global Turbulence?*

Dr Stefan Fritsch

Technology is neither good, nor bad, nor is it neutral.[1]

7.1 Introduction

Technology, defined as the accumulation of knowledge and artefacts for realizing human objectives in specifiable and reproducible ways,[2] has always played a vital, if not central, role in international relations (IR) or international political economy (IPE). The history of the human race offers countless examples of this with regard to military, economic, social and cultural developments.[3] The most profound effect of technological progress, especially since the fifteenth century, has been an increased density of the international system, caused by increasing and more rapid interaction. During the twentieth century, interaction capacities culminated in two important technological inventions: (1) nuclear weapons of mass destruction and their carrier systems; and (2) electrical and, later, electronic information and communication technologies (ICTs). The development of ICTs started in the second half of the nineteenth century (telegraph and telephone). Since the 1940s ICTs have spread to areas such as microelectronics and, computer technology (hardware), as well as related software development. Processes of convergence with telecommunication technologies and optoelectronics since the 1970s, enabled by the basic technological principle of digital binary code, have formed the basis for fundamental transformations in transnational politics, communication, finance, trade, production and culture.

---

* This chapter is a short extract from a PhD research project at the Department of History and Political Science, University of Salzburg/Austria, which is sponsored financially by the Doctoral Scholarship (DOC) Programme of the Austrian Academy of Sciences.

These two examples not only illustrate the significant influence of technology on military, political, economic and social developments, but also represent another quality of technology – namely its potential to transcend state borders, thereby undermining territorial sovereignty and authority, which have been the basic principles of the modern international system of nation-states as it developed after the Westphalia settlement of 1648. This argument was true for nuclear weapons during the Cold War[4] and it is true for ICTs in the era of the so-called 'information society'.[5] There exist a large number of scientific studies concerned with questions of specific technologies and their impact on state policies or the technologically motivated emergence of new policy issues.[6] Despite this, IR/IPE theory to date has been unable to formulate a clear and far-reaching theory, which gives sufficient weight to science and technology and its role in international affairs. With regard to this 'gap' Skolnikoff concludes: 'Even scholars concerned with theoretical issues in international relations tend to treat science and technology as static "givens", or as emanating from impenetrable black boxes'.[7]

The goal of this short chapter is to show how IR/IPE theorists have tried, over the course of the last few decades, to describe and analyse the growing role of technology (especially network-based ICTs) in shaping the political and economic structures and processes within the international system. Since it would be impossible to analyse a larger number of approaches, this chapter focuses on three central theoretical approaches that have had significant influence on IR/IPE theory during the last three decades: realism/neorealism, interdependent globalism and constructivism. The author generally argues that technology, in most cases, is still an undervalued (dependent) parameter of IR/IPE theory. It concludes that if the disciplines of IR and IPE want to analyse the influence of technological progress on international relations more effectively, then they will have to change some of their basic assumptions.

7.2 Realistic and neorealistic approaches to technology

With its basic assumptions of an international system of states (as the only important actors), realism defines IR as a war of all against all. According to realist understanding, states are primarily concerned with the pursuit of 'national interest', which is the maximization of power to ensure state security and survival in an anarchical system (one which is characterized by a lack of global governance) and the maintenance of the balance of power to evade the development of stronger opponents.[8] In these realist conceptions, technology had a 'black box' status. For realists technology was, and mostly still remains, a mere instrument of power to realize state goals.

Neorealism emphasizes the anarchic nature of a global society without governance rather than the traditional realist emphasis on the unceasing lust for power inherent in human nature in its explanation for why inter-state conflict persists. For this reason, neorealism is sometimes referred to as structural realism because it emphasizes the influence of the global power structure on the behaviour of states within it. Neorealism performed little better than realism in paying attention to technology. Kenneth Waltz, one of the most influential representatives of neorealism, paid close attention to the power distribution between states, as the basic factor for explaining the systems' character: 'A systems theory requires one to define structures partly by the distribution of capabilities across units. States, because they are in a self-help system, have to use their combined *capabilities* in order to serve their interests. [...] Their rank depends on how they score on all of the following items: size of population and territory, resource endowment, economic capability, military strength, political stability and competence'.[9] Although technology itself is not mentioned explicitly, it becomes obvious that it is seen as an implicit capability.

With these positions in mind, it can be said that both realism and neorealism display a rather instrumental understanding of technology. In these frameworths, technology represents a neutral tool, which has to be used to: (1) secure one's (in fact the state's) power position; or (2) to realize absolute or relative gains in relation to competitors (in areas of security, power or welfare). This non-conception of technology can also be found in contemporary (neo)realist works.[10] Generally, realism/neorealism showed remarkable ignorance of the fact that some modern technologies, like nuclear weapons or ICTs, can and do exert transnational and system-wide (re)structuring power, narrowing the room for states' sovereign political action and thereby displaying qualities of at least partial autonomy. As soon as realistic/neorealistic authors start to pay attention to technologically induced transformation processes, they have to reassess some of their most central assumptions.[11]

7.3   Interdependent globalism

The end of the Cold War fostered intensified economic and political exchange processes in global scope as well as the emergence of new powerful actors such as multinational corporations (MNCs), inter-governmental organizations (IGOs), nongovernmental organizations (NGOs) or even individuals. Modified hierarchies in the international system and the gradual loss of state power led many theorists of IR/IPE to question the traditional approaches of IR/IPE. States are no longer regarded as the only relevant actors on the world stage, but are confronted with sometimes equally empowered non-state actors. Many theorists consider ICTs to be one of the main driving forces behind these fundamental shifts. The newly arising global structures led Rosenau, one of the most prominent representatives of this new perspective in IR/IPE, to coin the new term of post-international politics:

The very notion of 'international relations' seems obsolete in the face of a apparent trend in which more and more of the interactions that sustain world politics unfold without the direct involvement of nations or states. [...] Postinternational politics is an appropriate designation because it clearly suggests a decline of long-standing patterns without at the same time indicating where the changes may be leading. It suggests flux and transitions even as it implies the presence and functioning of stable structures. It reminds us that 'international' matters may no longer be the dominant dimension of global life, or at least that other dimensions have emerged to challenge or offset the interactions of nation-states.[12]

What causes this systemic shift? According to Rosenau, the sources of these transformations is 'turbulence', which he defines as 'a world wide state of affairs in which the interconnections that sustain the primary parameters of world politics are marked by extensive complexity and variability'.[13] Rosenau also formulates a restriction to his model. States may lose some of their power to influence structures or processes, but they do not become obsolete:

States are changing, but they are not disappearing. State sovereignty has been eroded, but it is still rigorously asserted. Governments are weaker, but they can still throw their weight around. [...] Borders still keep out intruders, but they are also more porous. Landscapes are giving way to ethnoscapes, mediascapes, ideoscapes, technoscapes, and finanscapes, but territoriality is still a central preoccupation for many people.[14]

What are the sources of this relative power loss, according to those authors who stress the systemic transformations? Analysing the relevant literature reveals several interconnected and mutually reinforcing parameters, which play a central role: (1) a quantitative increase in the number of traditional actors (states and IGOs) as well as the emergence of qualitatively new actors (NGOs, MNCs, individuals),[15] (2) various globalization processes, namely in military, economic and socio-cultural areas[16] and (3) technology as one of the central sources for global turbulence.[17] Francis Fukuyama has probably formulated the most far-reaching globalistic position related to technology. For him, technological progress is the source for fundamental transformation that will result in the 'end of history'. By that, he understands the globally triumphant spreading of democracy and the liberal market economy. At the same time, these two concepts offer the most effective starting point for fostering these techno-economic developments.[18] Fukuyama is convinced that more and more states and their societies will accept these demands on their way to higher living standards.[19]

Although interdependent globalism has often been criticized as being too deterministic,[20] meaning that it describes globalization forces (mainly technology) as the dominant source for eroding states' policy formulation and implementation capacities, most representatives of this approach still see technology as a nondeterministic and external factor:

> Without new transportation and communications technologies, the dynamics of change presently at work would not be unfolding; but this is not a form of *determinism* because the new technologies do not carry humankind in a *single direction*. They facilitate several causal streams, and it is the dynamics of individuals and communities that determine which stream will be followed in particular situations or regions.[21]

7.4   Technology and IR/IPE from a constructivist point of view

Since the middle of the 1980s IR theory has been enriched by another perspective on international relations – one which, since its appearance, has caused a lot of discussion. In the words of Onuf: 'Constructivism is a way of studying social relations – any kind of social relations.'[22] Although constructivist approaches vary widely in their perception of social relations and processes, two aspects form a minimum denominator: (1) knowledge and meaning as well as (2) the idea that social reality are socially constructed. All constructivists try '[...] to "denaturalize" the social world, that is to empirically discover and reveal how the institutions and practices and identities that people take as natural, given, or matter of fact, are, in fact, the product of human agency, of social construction'.[23] Constructivists criticize realism/neorealism and globalism as materialist theories. For Wendt, one of the main representatives of moderate constructivism, a theory gains materialist status when '[...] it accounts for the effects of power, interests, or institutions by reference to "brute" material forces – things which exist and have certain causal powers, independent of ideas, like human nature, the physical environment, and perhaps, *technological artefacts*.'[24]

For Wendt, '[M]aterial capabilities as such explain nothing; their effects presuppose structures of shared knowledge, which vary and which are not reducible to capabilities.'[25] Technology interacts with society and culture, thereby influencing social behaviour and societies in certain ways: 'The term "interaction" is significant here, since it means that at some level material forces are constituted independent of society, and affect society in a *causal* way. Material forces are not constituted solely by social meanings and social meanings are not immune to material effects.'[26] The constructivist position on technology could be summed up in the following way: Technology as material structure influences political, social and economic processes. The mere existence of technology transforms social reality, but its actual meaning depends on the social context.[27] Nevertheless, constructivists tend to overestimate the power of ideas and identities in international relations and often underestimate the determining role of technology on the actors in the international system.

The conceptions of technology in IR/IPE theory that have been presented so far clearly show a growing concern for technology's role in international affairs. Nevertheless, they all fall short of paying in-depth attention to explaining the basic relationship between technology and the evolution of international affairs. Therefore, it seems necessary to combine IR/IPE theory with results of another strand of social research – namely the history, sociology and philosophy of technology.

7.5   Arguments for a broader perspective on technology

The questions in the neighbouring disciplines mentioned above revolve around the following problem: Is technology a tool that pre-exists and is simply used by society according to political and economic demands? Or, is technology an independent source of change, which itself creates political qualities, thereby predetermining the ways in which politics have to react, economies have to change and societies have to adapt? The first approach to technology, emphasizing its passive quality, and thereby enabling society to use it according to its interests, has been described as the social constructivist understanding of technology.[28] This approach represents a rather instrumental understanding of technology, according to which society can use technology according to its goals, being able to control all possible negative and positive side effects. In such a reading, technology is not understood as having political qualities per se – it is basically a neutral tool, a means to an end.[29] The opposite point of view has generally been associated with technological determinism. Authors like Lewis Mumford, Jacques Ellul or Langdon Winner tried to show that technology could gain political qualities as a partially independent force. Technological options might be actively chosen by society, but, in turn, they develop pressures for different actors to adapt, thereby forming political, economic and social processes, which might not always have been expected or deliberately chosen.[30]

These approaches have been under strong criticism, being described, in Marxian terms, as 'technological determinism', leaving out the possibility of reversing unforeseen developments by modifying technologies. Nevertheless, as the current author believes, some of their assumptions in order to allow IR/IPE theory to understand better the ways in which technology interacts with international politics, economics and global society. Taking into consideration the fact that states are no longer capable of managing technological, economic and social developments independent of one another, at least some deterministic arguments (loss of states' sovereignty to decide and implement independently) seem to be confirmed. Nevertheless, growing global efforts to solve many technology-related problems within various international institutions also show that new possibilities for influence and regulation of global developments do exist. This thereby opens up the arena for arguments of the social constructivist approach, which assumes the basic possibility of sustained control of technological developments, this time at a global level.

7.6 Multidimensional effects of technology

7.6.1 Individual level

The term 'skill revolution' describes the growing skills of individuals in analysing the manifold processes in global economics and politics, as well as social and cultural interaction processes. ICTs have been the basis for this improved information-gathering and -analysing capacity. The geographical distribution over large parts of the globe, as well as an ongoing decline in costs to buy and use them, have resulted in a 'democratization of technology', making it available for most people – at least in the industrialized world.[31] Moreover, individuals have also learned how to articulate their interests and how to network with other people who share the same ideas, identities or enemy images simply by using ICTs like e-mail or mobile phones. This enables individuals to form new collectives to articulate ideas and interests and/or implement them on a global scale.[32] One example of this skill revolution was the Zapatista's Internet information campaign against trade liberalization within the North American Free Trade Area in 1992, which they perceived to be endangering their traditional lifestyle by increasing cheap foreign grain imports.[33] Another instance was the broad anti-globalization movement. Another example of enhanced coordination capacities, supported by ICTs and modern transportation technologies like airplanes, are the terrorist attacks of September 11 2001 on New York and Washington. The growing dependence of modern societies on a wide range of technologies, as has been shown in many contributions in this book, opens up new potential targets for all kinds of terrorism.[34] In the economic realm, individuals can gain more autonomy of decision through the Internet by informational arbitrage (for example compare the price for a certain commodity on different websites), thereby improving the efficiency of their consumer decisions.[35] Brown and Studemeister come to the following conclusion concerning the growing influence of individuals in international relations: 'Single states find it difficult, not to say impossible, to respond effectively to such supra-individuals as Bill Gates, George Soros, Jimmy Carter or Osama bin Laden'.[36]

7.6.2 New structures in world politics

This point refers to the partial erosion of state sovereignty through different globalization and/or transnationalization processes, primarily driven by ICTs and transportation technologies. In economics, the pressure exerted on states by financial markets or MNCs, to constantly adapt to demands of global capitalism, has led some authors to speak of the emergence of new state qualities, such as the 'trading state',[37] the 'competition state'[38] or the 'privatization and marketization of the political-economic structure', which place the welfare state of the industrial era under severe strain.[39] States have to adapt to these new circumstances if they want to rescue at least some of their power. In the economic realm, with special regard to information-intensive and costly high technology, this means constantly improving national systems of education and research, assuring attractive milieus for foreign direct investment, the creation of highly skilled and flexible national workforces, up-to-date technological infrastructure (ICTs and transportation) and so on.[40] Of course, states can still influence markets and technological developments and try to control information flows. The US government, for example, played an important role in the global spread of the Internet and the liberalization of once state-owned communication monopolies through the WTO (World Trade Organization) for the profit of its own MNCs.[41] But the state's role here relates to the creation and support of general global structures for economic action. This does not prevent the USA, like other western industrialized states, from processes like outsourcing and offshoring, a hotly contested economic issue during the last presidential elections. The control of information flows will never be complete. In reaction to the stronger control of internet traffic and mobile communication, terrorists bypass these ICTs by older communication modes like videotapes and couriers.

MNCs, especially those in high-technology sectors, are also confronted with intensified global competition for leadership in technology and market shares. They constantly have to adapt to new technological standards and consumer demands. Since product cycles become shorter and shorter, and research and development (R&D) of new products becomes more expensive, many MNCs initiate strategic partnerships with competitors to reduce costs for R&D and/or try to expand into new markets to amortize the R&D expenses. Another very important structural aspect of global technology sectors has been the question of technological standards. This is probably one of the most vital aspects for economic success in general. The best example might be the now strongly emerging open source movement in the software sector, which has been heavily opposed by major established players like Microsoft.[42] Another example for the growing importance of technological standards is the ongoing conflict over the technological standard for the next generation of wireless mobile communication between European, US and East Asian communication equipment producers.

States, markets and MNCs are not the only actors in the international system that have gained significant potential to act globally. New social actors like NGOs, social movements or subgroups, generally defined as representatives of civil society, have also extended their skills. Many of these actors themselves make intensive use of ICTs. Litfin has shown that NGOs concerned with deforestation or global warming use high-resolution photos, provided by civil commercial satellite companies, to raise concern for these issues and thereby put pressure on state administrations to react.[43] Anti-globalization groups, anti-landmine groups, and so on, coordinate their activities and seek global support through the Internet and mobile communication devices.[44] Some NGOs that are concerned with the continuing freedom of the Internet itself, such as Privacy International, spread their messages via cyberspace and try to raise concern for the ongoing commercialization of the Internet by measures of intellectual property protection, forcefully driven by the new MNCs of the twenty-first century.[45]

7.6.3   New modes of interaction

All of the processes described above lead to new modes of governance in the international system. States are no longer in an exclusive position to 'dictate' the rules of the political and/or economic game. They react either by increasing cooperation through international governmental institutions (IGOs or regimes) or by integrating into larger polities or economic regions – for example, the European Union.[46] This provides them with the possibility of gaining new power to shape basic political and economic conditions for future technological developments.[47] As van Creveld has noted, states who want to gain access to global technological networks have no other possibility but to join those international bodies '[...] whose task [is] to regulate the technologies in question on behalf of all'.[48] At the same time, they must intensify cooperation with non-state actors like MNCs, NGOs or other subgroups that rival states at least in some aspects of IR/IPE.[49] Rosenau has described the emergence of these new fields of interactions between political, social and economic non-state actors who develop their own – private – authority (based on values, norms, interests, local origin or orientation) as the 'multicentric world' with new 'spheres of authority' (SOAs).[50] This multicentric world of global economy, global civil society, countless local interest groups, and so on, constantly interacts with the still existing classical state-centric world. It consists of polities with declining authority in many different ways, creating new forms of 'governance without government', which means a complex web of many different actors, interacting with one another on different levels to gain influence or regulate problem areas.[51]

7.7   Conclusion

Returning to the opening statement by Kranzberg – that technology is neither good, nor bad, nor neutral – it should have become clear from this chapter that technology in IR/IPE represents more than a passive tool or a pre-existing black box, but at the same time is less than a completely independent 'technological Frankenstein-technology out of control of human agency'.[52] Technology has a deep influence on actors, their identities and interests as well as on the international system's structures and processes. On the individual level, technology might be used in different ways: The Internet, for example, can be used to do business, to communicate with friends, or to organize and carry out terrorist attacks. States and MNCs also make decisions concerning the development and distribution of new technologies, without being able to assess all possible political, social and economic (side-)effects of these technologies. Finally, on a global level, technology in most of its contemporary applications tends to develop into mega-systems, embracing most parts of the world and acting at least partly autonomous (for example negative side-effects, the predetermination of future (technological) decisions, and so on).[53] It is on a global level that technological systems create partly deterministic qualities, understood as a certain dependence of the human race on technology. The reader may just try to imagine a world without public transportation, cars, airplanes, ICTs, and so on. What would be the effects? In many cases, technology's future development, benefiting as many people as possible, can only be (re)regulated on a global level, increasing the need for global governance efforts between all system actors.

Only when theories of IR/IPE seek to pay more attention to the discussions led within neighbouring social science disciplines with regard to technology will they be able to gain additional insights in the ways it influences the international system, its structures, its processes and its actors. To sum up the overall perception, in the history of IR/IPE theory technology has been gradually gaining importance as an analytical category to describe, explain and predict political, economic and social developments in the international system. Nevertheless, technology still appears to be an undervalued passive as well as exogenous factor in most mainstream theoretical frameworks presented here. A lot of future research needs to be done within IR/IPE theory in order to properly integrate technology and its growing role in the theories of IR/IPE.

Notes

1. M. Kranzbert, 'The Information Age: Evolution or Revolution?' in B.R. Guile (ed.), *Information Technologies and Social Transformation* (Washington DC: National Academy of Engineering, 1985), p. 50.
2. H. Brooks, Technology, Evolution and Purpose in Modern Technology. *Daedalus*, 109(1) (1980), 65–81.
3. For military technological developments see M. van Creveld, *Technology and War: From 2000 B.C. to the Present* (New York: The Free Press, 1989); W.H. McNeill, *The Pursuit of Power: Technology, Armed Forces, and Society since A.D. 1000* (Chicago: University of Chicago Press, 1982); and A. Pacey, *Technology in World Civilization:*

A Thousand-Year History (Cambridge, MA: MIT Press, 1990). For the role of science and technology in western economic development, see N. Rosenberg and L.E. Birdzell, *How the West Grew Rich: The Economic Transformation of the Industrial World* (New York: Basic Books, 1986); and D. Landes, *The Wealth and Poverty of Nations: Why Some are so Rich and Some so Poor* (New York: W.W. Norton, 1998).

4. See especially the programmatic article by B. Brodie, 'The Atom Bomb as Policy Maker', *Foreign Affairs*, 27(1) (1948/49), 17–33. Also see R. Jervis, 'The Political Effects of Nuclear Weapons: A Comment', *International Security*, 13(2) (1988), pp. 80–90; and C.S. Gray, *The Geopolitics of the Nuclear Era: Heartlands, Rimlands, and the Technological Revolution* (New York: Crane, Russak & Company Inc., 1977).

5. W. Wriston, 'Technology and Sovereignty', *Foreign Affairs*, 67(2) (1988/89), 63–75.

6. For probably one of the earliest attempts to analyse different technologies and their impact on IR see W.F. Ogburn (ed.), *Technology and International Relations* (Chicago: The University of Chicago Press, 1949).

7. E.B. Skolnikoff, *The Elusive Transformation: Science, Technology and the Evolution of International Politics* (Princeton, NJ: Princeton University Press, 1993) p. 9.

8. As important representatives of this approach see: H.J. Morgenthau, *Politics among Nations* (New York: Knopf, 1948); and R. Niebuhr, *Moral Man and Immoral Society* (New York: Scribner's, 1947).

9. K.N. Waltz, *Theory of International Politics* (New York: McGraw-Hill Inc., 1979) p. 131. Emphasis added by the author. The book's index does not even list the term 'technology'.

10. Z. Brzezinski, *American Primary and Its Geostrategic Imperatives* (New York: Basic Books, 1997).

11. Three examples may prove this assumption:

   1. In 1961, Hans Morgenthau formulated with regard to nuclear weapons a somehow modified position: 'I think a revolution has occurred, perhaps the first true revolution in foreign policy since the beginning of history, through the introduction of nuclear weapons into the arsenal of warfare. For from the beginning of history to the end of the Second World War, there existed a rational relationship between violence as a means of foreign policy, and the ends of foreign policy. That is to say, a statesman could ask himself – and always did ask himself – whether he could achieve what he sought for his nation by peaceful diplomatic means or whether he had to resort to war. [...] the statesman in the pre-nuclear age was very much in the position of a gambler [...] who is willing to risk a certain fraction of his material and human resources. If he wins, his risk is justified by victory; if he loses, he has not lost everything. His losses, in other words, are bearable. This rational relationship between violence as a means of foreign policy and the ends of foreign policy has been destroyed by the possibility of all-out nuclear war.' See H. Morgenthau, 'Western Values and Total War', *Commentary*, 32(4) (1961), 4, p. 280.

   2. In the same respect, Robert Gilpin comments: '[...] atomic weaponry and rockets have indeed had a significant impact on the nature and instruments of statecraft. [...] there may be some ground for cautious optimism in that the threat of thermonuclear weapons to national survival gives nations, for the first time in history, a strong overriding common interest in avoiding war. The pursuit of this common interest, therefore, in terms of formulating appropriate international institutions and codes of behaviour becomes one of the foremost

challenges modern technology poses for the statesman today [...].' See R. Gilpin, *Has Modern Technology Changed International Politics?*, in J.N. Rosenau, V. Davis, M.A. East (eds), *The Analysis of International Politics: Essays in Honour of Harold and Margaret Sprout* (New York: The Free Press, 1972) p. 173.
3. B. Buzan and R. Little, who develop their concept of 'interaction capacity' for differing between ancient, classical and modern international systems, formulate a modified neorealistic position. According to their point of view rapid advances in modern military, transportation and communication technologies have greatly increased interaction capacities, thereby creating – for the first time in human history – a truly global international system. Furthermore, technological progress has brought forth new actors, who partly rival the classical state as dominant actor in the international system. See B. Buzan and R. Little, *International Systems in World History: Remaking the Study of International Relations* (Oxford: Oxford University Press, 2000).

To be fair, one has to note further works, generally associated with realism, who have explicitly paid attention to technology and its structuring effects on the international system and its actors: R. Aron, *Peace and War: A Theory of International Relations* (New York: Doubleday & Co., 1966); R. Gilpin, *War and Change in World Politics* (Cambridge: Cambridge University Press, 1981); and R. Gilpin, *The Political Economy of International Relations* (Princeton, NJ: Princeton University Press, 1987).

12. J.N. Rosenau, *Turbulence in World Politics: A Theory of Change and Continuity* (New York and London: Harvester-Wheatsheaf, 1990), p. 6.
13. Ibid., p. 78.
14. J.N. Rosenau, *Along the Domestic–Foreign Frontier: Exploring Governance in a Turbulent World* (Cambridge: Cambridge University Press, 1997) p. 4.
15. Rosenau, *Turbulence in World Politics* p. 12f.
16. D. Held, and A. McGrew, and D. Goldblatt, and J. Perraton, *Global Transformations. Politics, Economics and Culture* (Cambridge: Polity Press, 1999).
17. R. Langhorne, *The Coming of Globalization: Its Evolution and Contemporary Consequences* (Basingstoke: Palgrave Macmillan, 2001), p. 2.
18. F. Fukuyama, *The End of History and the Last Man* (New York: Perennial, 1993), pp. 71–108.
19. Ibid., p. 108.
20. B.R.J. Jones, *The World Turned Upside Down? Globalization and the Future of the State* (Manchester: Manchester University Press, 2000)
21. Rosenau, *Along the Domestic–Foreign Frontier*, p. 47. Emphasis added by the author.
22. N. Onuf, 'Constructivism: A User's Manual', in V. Kubalkova, N. Onuf and P. Kowert (eds), *International Relations in a Constructed World* (Armonk, NY: M.E. Sharpe, 1998), S. 58.
23. T. Hopf, 'The Promise of Constructivism in International Relations Theory' *International Security*, 23(1) (1998), 182.
24. A. Wendt, *Social Theory of International Politics* (Cambridge: Cambridge University Press, 1999) p. 92.
25. Ibid., p. 73.
26. Ibid., p. 111.
27. J. Checkel, 'The Constructivist Turn in International Relations Theory', *World Politics*, Vol. 50 (1998), no. 1, p. 326.

28. P. Weingart (ed.), Technik als sozialer Prozeß (Frankfurt am Main: Fischer, 1989). W.E. Bijker, T.P. Hughes and T.J. Pinch, (eds), *The Social Construction of Technological Systems: New Directions in the Sociology and History of Technology* (Cambridge, MA: MIT Press, 1987).

29. Literature generally differentiates between radical and moderate sociocentric approaches. For an overview, see N.J. Vig, 'Technology, Philosophy, and the State: An Overview', in M. Kraft and N.J. Vig (eds), *Technology and Politics* (Durham and London: Duke University Press, 1988), pp. 8–32. A recently published introductory work to the changing philosophy of technology is: D.M. Kaplan (ed.), *Readings in the Philosophy of Technology* (Lanham, Boulder et al.: Rowman & Littlefield Publsihers Inc., 2004). See also U. Teusch, *Freiheit und Sachzwang. Untersuchungen zum Verhältnis von Technik, Gesellschaft und Politik* (Baden-Baden: Nomos Verlagsgesellschaft, 1993).

30. L. Mumford, *Technics and Civilization* (New York: Harcourt Brace & Company, 1934); J. Ellul, *La Technique ou L'enjeu du Siècle* (Paris: Economica, 1990 [1964]); and L. Winner, *Autonomous Technology. Technics-out-of-Control as a Theme in Political Thought* (Cambridge, MA: MIT Press, 1977). For an in-depth analysis of technological determinism see L. Marx and M.R. Smith (eds), *Does Technology drive History? The Dilemma of Technological Determinism* (Cambridge, MA: MIT Press, 1994).

31. T.L Friedmann, *The Lexus and the Olive Tree* (New York: Farrar, Straus & Giroux, 1999).

32. J.N. Rosenau, and M.W. Fagen, 'A New Dynamism in World Politics. Increasingly Skillful Individuals?', *International Studies Quarterly*, 41(4) (1997), 655–86.

33. The Zapatistas' use of the Internet has generally been cited as one of the first examples for new forms of conflict like virtual netwar. See D. Ronfeldt and A. Martinez, 'A Comment on the Zapatista "Netwar"', in J. Arquila and D. Ronfeldt (eds), *In Athena's Camp: Preparing for Conflict in the Information Age* (Rand: National Defence Research Institute, 1997), pp. 369–391.

34. T. Homer-Dixon, 'The Rise of Complex Terrorism', *Foreign Policy*, 1 (2002), 52–62.

35. K. Ohmae, *Der unsichtbare Kontinent: Vier strategische Imperative für die New Economy* (Vienna: Ueberreuter Verlag, 2001).

36. S.J. Brown and M.S. Studemeister, *Diffusion of Diplomacy. Net Diplomacy I. Beyond Foreign Ministries, Part III*. United States Institute of Peace 2002. Available at www.usip.org/virtualdiplomacy/ publications/reports/14c. html.

37. R. Rosecrance, *The Rise of the Trading State* (New York: Basic Books, 1986).

38. P.G. Cerny, *The Changing Architecture of Politics: Structure, Agency, and the Future of the State* (London: Sage Publications, 1990).

39. Ibid., p. 339.

40. For a detailed introduction to these issues, see R. Palan, J. Abbott, and P. Deans, *State Strategies in the Global Political Economy* (London and New York: Pinter, 1996).

41. D. Schiller, *Digital Capitalism: Networking the Global Market System* (Cambridge, MA: MIT Press, 1999).

42. T. Baumgärtel, 'Am Anfang war alle Software frei. Microsoft, Linux und die Rache der Hacker', in A. Roesler and B. Stiegler (eds), *Microsoft: Medien, Macht, Monopol* (Frankfurt am Main: Edition Suhrkamp, 2002), pp. 103–29. See also K. Sangbae and J.A. Hart, 'The Global Political Economy of Wintelism: A New Mode of Power and Governance in the Global Computer Industry,' in J.N. Rosenau, and J.P. Singh (eds), *Information Technologies and Global Politics: The Changing Scope of Power and Governance* (Albany: State University of New York Press, 2002), pp. 143–68.

43. K. T. Litfin, 'Public Eyes. Satellite Imagery, The Globalization of Transparency, and New Networks of Surveillance', in J.N. Rosenau and J.P. Singh (eds), *Information Technologies and Global Politics: The Changing Scope of Power and Governance* (Albany: State University of New York Press, 2002), pp. 65–89.

44. For some detailed examples of NGOs and their use of ICTs see C. Warkentin, *Reshaping World Politics: NGOs, the Internet, and Global Civil Society* (Lanham, MD and Boulder, CO: Rowman & Littlefield Publishers Inc., 2001).

45. For a critical assessment of these problematic developments, see L. Lessig, *Code and Other Laws of Cyberspace* (New York: Basic Books, 1999) and L. Lessig, *The Future of Ideas: The Fate of the Commons in a Connected World* (New York: Random House, 2001).

46. P.G. Cerny, 'Globalization and the Changing Logic of Collective Action', *International Organization*, 49(3) (1995), 595–625. For economic regionalization, see K. Ohmae, 'The Rise of the Region State', *Foreign Affairs*, 72(2) (1993), 78–87.

47. One (supranational) example would be the efforts within the European Union to develop strategies for a successful participation in the global information economy. For these efforts see: European Union, *eEurope 2002: An Information Society for All* (Action Plan prepared by the Council and the European Commission for the Feira European Council, 19–20 June 2000).

    Another example for efforts within IGOs to solve global problems arising with the global information society (for instance, the growing digital gap between the information rich and the information poor) would be the ongoing 'World Summit on the Information Society' which is organized by the International Telecommunication Union (a special organisation of United Nations, situated in Geneva/ Switzerland) between early 2003 and 2005 in Geneva and Tunis, Tunisia. This process aims to bridge the 'digital divide' between the North and the South.

48. M. van Creveld, *The Rise and Decline of the State* (Cambridge: Cambridge University Press, 1999) p. 380.

49. One example for the cooperation of MNCs and states would be the case of an Ericsson-led European consortium that persuaded the European Union to adopt the GSM standard for mobile communication, which was not backward compatible with second-generation standards of the US corporation Qualcomm, thereby threatening the long-run viability of Qualcomm. After intense high-level controversy, Ericsson bought Qualcomm's network business and an agreement was struck to support two standards. Another example for such state–MNC alliances are the negotiations within the ITU between communication equipment producers, service providers and states for the new global mobile communication standard IMT-2000.

50. Rosenau, p. 15. Ferguson and Mansbach come to similar conclusions, but use the more familiar term 'polities'. See Y.H. Ferguson and R.W. Mansbach, 'History's Revenge and Future Shocks. The Remapping of Global Politics', in M. Hewson and T.J. Sinclair (eds), *Approaches to Global Governance Theory* (Albany: State University of New York Press, 1999), pp. 197–238. Castells uses the term 'network society' to describe the same results. See M. Castells, *The Information Age: Economy, Society and Culture Vol. 1: The Rise of the Network Society* (Basic Books: New York, 2000).

51. J.N. Rosenau, 'Governance, Order, and Change in World Politics', in J.N. Rosenau, and E.O. Czempiel (eds), *Governance without Government: Order and Change in World Politics* (Cambridge: Cambridge University Press, 1992), p. 4.

52. J.P. Singh, 'Introduction. Information Technologies and the Changing Scope of Global Power and Governance', in J.N. Rosenau and J.P. Singh (eds), *Information*

112 *Cyberwar, Netwar and the Revolution in Military Affairs*

Technologies and Global Politics: The Changing Scope of Power and Governance (Albany: State University of New York Press, 2002) p. 11.

53. T.P. Hughes, 'Technological Momentum', in L. Marx and M.R. Smith (eds), *Does Technology Drive History? The Dilemma of Technological Determinism* (Cambridge, MA: MIT Press, 1994), pp. 101–13.

8
Nuclear Weapons and the Vision of Command and Control

Dr Bruce D. Larkin

Washington blundered into Iraq in 2003 and, bungling the aftermath, could not see a clear exit. The 'lone superpower', the 'hyperpower', the military master of the universe, has simply made the wrong decision. Moreover, its failures are self-inflicted. The White House and civilian Pentagon leadership persuaded themselves of a grand fantasy. They neglected – ignored, or spurned – intelligence fundamentals. They claimed as facts what they could not prove. They assumed all would go as they wished, and that their forces would be greeted as liberators. They had no doubt that US military power would prevail.

This chapter is about nuclear weapons, rather than the Iraq War of 2003 and its follow-up, but it does address a question posed by the war. If Washington could get so much wrong in invading Iraq, can we be confident of its management of nuclear weapons?

War policy and nuclear policy are inseparable, even if one dismisses as 'out of the question' the possibility of nuclear use in *this* war. The commander-in-chief who chose to war in Iraq also commands the US nuclear forces. He is half of the National Command Authority (NCA), which, in principle, must authorize any nuclear weapon use, and is also commander-in-chief of the US military. The other half of the NCA, the secretary of defense, is the very same Donald Rumsfeld who advocated attacking Iraq as an appropriate response to the September 11 attacks, and who has overseen the conduct of the Iraq War. The January 2002 Nuclear Posture Review, which calls for new nuclear weapon designs and steps to make the resumption of nuclear testing easier, and the September 2002 report on National Security Strategy, making a case for US 'pre-emption' in the face of 'imminent threats', advance White House and Pentagon arguments for new military capabilities and state their readiness to use them.

Of course, in such short chapter I cannot consider every aspect of 'the management of nuclear weapons'. This chapter explores nuclear decision and execution in the USA by focusing on the access of the leadership and the military to prompt, informative, and secure communications.

Managers intend, to manage nuclear weapons well and safely, and they have thoughtful procedures in place to do so. However, they are technologically constrained, open to misjudgements, capable of error, and dependent both on complex systems that may fail, and on assumptions about the loyalties and performance of their personnel. What works without disabling fault in routine and peacetime conditions may not work in the face of the unexpected circumstances of crisis decision and contemplated nuclear use. This means that we cannot be confident of the current management of US nuclear weapons. The better posture is one of scepticism towards claims that any system of command and control can meet the standards of coherence, robustness, reliability and performance which ongoing 'nuclear weapons management' requires.

8.1 The White House and the Department of Defense (DoD)

Nuclear weapons impose new requirements. The government of a nuclear power navigates between the imperative of restraint and fear of 'surprise attack'. Committed to deterrence or *dissuasion*, it deploys ready forces, but must prevent unintended and unauthorized use. If it chooses to keep open options for 'launch on warning' or 'launch under attack', it accepts a disappearingly short time between the receipt of any warning and the order to fire. Every step – notice, deliberation, crisis restraint, execution – relies on communications among authorized officials and military personnel.

The White House has many communications desiderata, but the most compelling ones concern the need to govern the use of nuclear weapons. These needs are met by the White House Communications Agency (WHCA), an arm of the DoD, staffed by military personnel. WHCA is tasked to ensure that the president and vice-president enjoy secure communications at all times, whether they are in Washington, elsewhere in the USA, in transit, or abroad. The standard model has it that nuclear weapons cannot be used without the president's consent, registered by his using a special code. Put out of mind, for the moment, circumstances in which a president could not act, due to death or incapacity; and put out of mind, too, your suspicion that the authority to use nuclear weapons has been 'predelegated' to lower echelons should a president be unable to function. What communications must be in place?

The NCA is vested in the president and the secretary of defense. The norm of the 'two-man rule' requires that at each level in the chain of command there be agreement by at least two authorized persons that nuclear weapons are to be used. Below the NCA that would mean verifying that a conforming order has been issued, and at the highest level the agreement of president and secretary of defense, or those acting as the NCA in their place, for use.

So there must be a way in which the president and secretary of defense, or their stand-ins, can exchange views.[1]

This requires that both be provided always-up, secure communications. Here 'secure' means not only that their messages, conversation, or accompanying information not be comprehensible if intercepted, but also that it not be spoofed. If a stand-in acts in place of the principal, there must be assurance that the substitution is appropriate. Adequate security is sought by encryption and the use of special code words.

8.2   The White House Communications Agency

The White House Communications Agency (WHCA) operates not only in the White House, but also wherever the president, vice-president and the president's spouse travel. Its Washington Area Communications Command also supports the Secret Service and White House Military Office within the District of Columbia. Three commands support communications during travel, and there is another at the presidential retreat, Camp David.[2]

WHCA is the current form of an organization that has been repeatedly revised, reconfigured and re-equipped since its antecedent White House Signal Detachment was created in 1942. By the late 1990s it had some 800 personnel, and the average tour of duty was four years. In 1994 the Agency was 'growing rapidly and asking for still more people at a time when DoD was shrinking everything'. A deliberate internal redesign was carried out, 80 per cent complete by 1998, which included notions of participative management 'not compatible with the traditional style of military management, which continues to be a key characteristic of WHCA's culture'.[3]

How well-prepared was this agency when faced with an actual crisis? 'After the initial attacks on the World Trade Center and the Pentagon, the White House staff was literally waiting in line to use secure communication lines.'[4] Did the White House not have state-of-the-art equipment and capabilities? The officer commanding WHCA's Washington Area Communications Command wrote in early 2003 that '[o]verhauling the agency's aged and failing legacy computer systems has been an ongoing project' and that they 'successfully field[ed] encryption cards and computer systems to WHCA and WHMO [White House Military Office] staff members for sending secure email transmissions'. On 1 September 2002 they opened an operations centre 'modelled after the Defense Information Systems Agency's Global Network Operations and Security Center' to provide round-the-clock monitoring of deployed systems and asset; with the implication that such a capacity had not previously been in place.[5]

The current programme to upgrade WHCA capabilities, a 'sweeping technological transformation plan' dubbed the Pioneer Project,[6] reflects G.W. Bush's comment that: 'I get – when I'm down at Crawford, I'm in constant contact with our administration. We've got secure teleconferencing capacity there. And it's pretty good. It can be better. (Laughter.) It can be more real-time.'[7] To that end the Defense Information System Agency FY2005 Budget Estimate contained a line supporting 'operational development of a secure, survivable voice conferencing capability for the national and military leadership during crises', titled Presidential and National Voice Conferencing. Planned expenditure of $9.975 million in FY2005 was intended to provide:

> system engineering, planning, development, integration, installation and testing of new baseband (cryptographic and vocoder) equipment needed to provide survivable, near toll-quality voice conferencing capability for the President and other national/military leaders. This project funds the critical and essential engineering required to develop a new voice processing algorithm, as well as the development of new vocoder and cryptographic equipment by taking advantage of ongoing RDT&E efforts by another Defense component. These baseband devices will implement new technology capabilities such as multi-stream cryptography/vocoding and information technology capabilities such as baseband Ethernet interfaces supporting baseband Internet Protocol (IP) addressing. This project supports the Joint Staff's requirement to fully implement the recommended Advance Extreme High Frequency (AEHF) PNVC improvements no later than FY10 for all PNVC participants.[8]

Another way of interpreting this passage is that these capabilities will not be in place until 2009 at the earliest.

8.3   Crisis experience: the attempted assassination of President Ronald Reagan

On 30 March 1981 US President Ronald Reagan was shot in Washington and rushed to hospital. Senior officials gathered in the White House Situation Room, where, with the permission of those present, National Security Advisor Richard V. Allen made a tape recording of the exchanges around the table.[9] Allen has published a portion of the transcript, which provides an extraordinary insight into the communications issues, confusion and errors that bedevilled the Center on that day. The Control of US nuclear weapons was an explicit concern of several participants.

The results of that day should have provided a wake-up call to those entrusted with nuclear command and control, but it is not far-fetched to suspect that different, and equally disabling, frailties haunt today's system. Several key points emerge from Allen's report:

- Vice-President G.H.W. Bush was 'in the air' returning from Texas, with a so-called 'football' carrying nuclear weapon release codes, and did not have a secure communication link to the White House. Secretary of State

Alexander Haig seemed to believe that, with Reagan on the operating table, Bush did command the codes: 'the football is near the Vice President – so that's fine'.

- Nonetheless, they may have thought a nuclear command could issue from those at the White House. Allen and Haig realized there was no 'football' in the Situation Room. Allen volunteered: 'We should get one over here. We have a duplicate one here...There is one at the military aide's office. The football is in the closet...'
- The Strategic Air Command was put on 'standby alert', though the DEFCON readiness level was not raised. Haig and Allen were focused on the possibility of ordering the use of nuclear weapons: 'HAIG: Do we have a football here? Do we? ALLEN: Right here.'
- White House Counsellor Edwin Meese called from the hospital, asserting erroneously 'that "the national command authority" rested with [Secretary of Defense Caspar W] Weinberger'. Weinberger believed that Bush's absence from the White House removed Bush from the chain of 'National Command Authority': 'WEINBERGER:...until the Vice President actually arrives here, the command authority is what I have.' But Haig differed: 'HAIG: You'd better read the Constitution. WEINBERGER: What? HAIG (laughing): You'd better read the Constitution. We can get the Vice President any time we want.' [Of course, the Constitution says nothing about a 'National Command Authority.']

In the event, no nuclear crisis took place. Still, we learn that there were three 'footballs' (one with Reagan, one 'near' Bush, one in the military aide's White House closet) and also, it would appear, that no 'football' accompanied the secretary of defense. Allen and Haig seem to have assumed that the 'football' in the closet held the current authorization codes.

8.4 Crisis experience: The September 11 attack

When the September 11 2001 attacks on the World Trade Center (WTC) and Pentagon took place, the president and the secretary of state were both away from Washington. Secretary of State Colin Powell was in Lima, Peru, attending a meeting of the Organization of American States; G.W. Bush was in Sarasota, Florida. The president learned that a plane had crashed into the WTC, and at 8.55 a.m., shortly before entering an elementary school classroom, he spoke to National Security Advisor Condoleeza Rice, who was at the White House. At 9.03 a.m. an aircraft struck the second tower, indicating that the crashes were deliberate.

As a result of the public report of the September 11 Commission, we have an unusually detailed source about recent White House and Department of Defense responses to crisis.[10]

The FAA [Federal Aviation Administration], the White House and the Defense Department each initiated a multiagency teleconference before 9:30. Because none of these teleconferences – at least before 10:00 – included the right officials from both the FAA and Defense Department, none succeeded in meaningfully coordinating the military and FAA responses to the hijackings.[11]

For at least an hour none of the information in the White House videoconference was passed to the National Military Command Center (NMCC). Moreover, the Commission wrote that 'we do not know who from Defense participated, but we know that in the first hour none of the personnel involved in managing the crisis did'. A witness told the Commission that:

> [It] was almost like there were parallel decisionmaking processes going on; one was a voice conference orchestrated by the NMCC...and then there was the [White House video teleconference]...[I]n my mind they were competing venues for command and control and decisionmaking.[12]

The NMCC opened an 'air threat conference call' at 9.37 a.m. This lasted 8 hours, and included, from time to time, the president, the vice-president, and the secretary of defense. Also at 9.37 a.m. the vice-president, now in a tunnel under the White House, paused at a secure phone, and asked to speak to the president, 'but it took time for the call to be connected.' At that point G.W. Bush was en route to the airport. They spoke at about 9.45 a.m.[13] Secretary of Defense Donald Rumsfeld was giving a briefing at the time the third aircraft struck the Pentagon and went from the briefing to the stricken area. Only later, shortly before 10.30 a.m., did he join the teleconference.

In Lima, Secretary of State Powell, on learning of the attacks, called for the aircraft which would carry him on the seven-hour flight to Washington, 'with poor phone connections'.[14] Though a statutory member of the National Security Council:

> it was a long day for me, as I got in my plane and flew all the way back from Peru, unable to communicate with anybody in Washington until I arrived and joined the President in the White House with the other national security advisors to the President.[15]

Although the September 11 case confirms that teleconference capabilities exist and that the principals could speak to one another readily when in locations with the right equipment, they often were not. It also reveals a disconnect between the White House and DoD in the hour after the second attack, and that as of 9.46 a.m. Pentagon staff 'were still trying to locate Secretary Rumsfeld',[16] another statutory member of the National Security Council (NSC).

8.5 The Global Command and Control System (GCCS) (as defined by the DoD)

The Pentagon's operations, including its preparations for nuclear operations, rely heavily on a fabric of computer-based telecommunications, now familiarly labelled 'information technology'. In the latter 1990s the US DoD packaged changes in capabilities and doctrine under the rubric Revolution in Military Affairs (RMA), emphasizing information technology and 'joint operations' (those of more than one service arm, such as the navy and the army).

In turn, an information network was put in place to support these operations: the Global Command and Control System (GCCS). In principle, GCCS integrates hardware, software and practice. Hardware includes sensors, workstations, servers, fibre, satellite receivers and transmitters, storage and computing. Marshalling these new capabilities to acquire and usefully display information that was hitherto unattainable, GCCS would permit the military to identify appropriate actions and coordinate and sequence their performance. Computers are now omnipresent. The acronym C3I, 'Command, Control, Communications, and Intelligence', has now been superseded: 'The Joint Staff . . . is working aggressively towards the Department's goal of seamless Command and Control, Communications, Computers, Intelligence, Surveillance, and Reconnaissance (C4ISR) interoperability by fiscal year 2008.'[17]

The US government advertises GCCS in Pentagon argot:[18]

GCCS is the nation's premier system for the command and control of joint and coalition forces. It incorporates the force planning and readiness assessment applications required by battlefield commanders to effectively plan and execute military operations. Its Common Operational Picture correlates and fuses data from multiple sensors and intelligence sources to provide warfighters the situational awareness needed to be able to act and react decisively. It also provides an extensive suite of integrated office automation, messaging, and collaborative applications.

('Coalition forces' are those of the foreign countries cooperating with the USA; but just how to give non-US militaries access to GCCS remains an ongoing issue.) In the 1990s military planners determined that they would rely heavily on commercial off-the-shelf products, avoiding the long lead times and uncertain quality of in-house design and development.

Built upon the foundation provided by the Common Operating Environment (COE), GCCS incorporates the latest in commercial computer hardware, software, and communications technology. Through an innovative evolutionary acquisition strategy GCCS is able to rapidly and cost-effectively field new applications as requirements evolve and technology advances.

GCCS is fielded at over 625 sites worldwide, all networked via the DOD's classified private Intranet. It is designed and implemented to provide our nation's warfighters the information superiority required to prevail now, and well into the 21st century.

In other words, GCCS is a dedicated 'Internet', a private Intranet like those maintained by corporations with widely dispersed activities. It is not open to the public, being designed as a *secure* system, a system for secret communications.

The problem for developers of GCCS was to take a zoo of service-designed specialized computer systems and enable common access and use: 'interoperability'. Distinct systems had to be able to talk to one another while maintaining their original security and assuring that gateways between components were also secure. The Common Operating Environment (COE) defines that connection point.[19] For each of the separate components, networking software has had to be written, taking the pre-existing service component on one side and passing data to and from the COE on the other side. Moreover, as we will elaborate below, changes in both component systems and GCCS features have been ongoing. New hardware is introduced, old hardware withdrawn, and data rates and volumes increased. The second problem for designers, therefore, is how to keep a high-reliability 'production system' up and available, while accommodating upgrades to components and introducing upgrades to the integrative GCCS software.

GCCS sites must be able to communicate with one another in the real world. How can this be done while ensuring that the 'enemy' cannot listen to communications, spoof them, or interfere with them? The answer is a dedicated network of *nested* capabilities, each implementing yet further security precautions. As the Defense Information System Agency (DISA) account reports, the broad system is: (1) the Defense Information System Network (DISN); then (2) the subset of DISN in which GCCS runs, the Secret Internet Protocol Router Network (SIPRNet), an IP network like our everyday Internet, and (3) a further subset of SIPRNet with additional encryption, the 'Top Secret (TS) version of GCCS (GCCS-T)'.

8.6   GCCS-T: the top secret provision for nuclear operations

This innermost, highly-protected, 'virtual private network' was commissioned to bring *nuclear weapons* planning, command, and control within the GCCS framework:

> Connectivity between all GCCS sites is provided through the Defense Information System Network (DISN). GCCS, operating in a Secret High security environment, is connected via the Secret Internet Protocol Router Network (SIPRNet) subset of the DISN. The Top Secret (TS) version of

GCCS (GCCS-T) is also connected via the SIPRNet, but use of a Network Encryption System (NES) in between the nodes creates a Virtual Private network (VPN) allowing TS to run over a secret network.[20]

In early 1998 the Pentagon declared that by mid-1998 it would integrate nuclear operations, the Single Integrated Operational Plan (SIOP), the National Command Authority's options for executing a nuclear attack, into GCCS-T:

> COMMAND AND CONTROL
> Command and control (C2) systems provide the means to effectively execute nuclear, conventional, and special operations...GCCS Top Secret (GCCS-T) provides a top secret infrastructure for C2 throughout the force deployment cycle. When completed in mid-1998, GCCS-T Version 2.2 will add nuclear Single Integrated Operational Plan capabilities and a top secret (including special intelligence) common operational picture. GCCS and GCCS-T improvements in 1999 will further add sensitive compartmented information, increase user sites, and improve perform-ance and reliability...[21]

Was 'will add nuclear SIOP capabilities' a phrase of art? Could they have meant, in 1998, only *some* capabilities, such as access to the most recently completed SIOP, but not the actual capacity to 'plan and execute'?

8.7 Ongoing transformation of command and control systems

By the mid-1990s, no longer preoccupied by Cold War concerns with Soviet nuclear forces, the US military had become focused on developing its C3I capabilities for 'conventional' tasks. War would not be fought as it had been in the past. Now it would take advantage of technological change, and espe-cially information technology (IT). In a series of documents, the US DoD spelled out the direction in which it would take US capabilities.[22] The US 1998 Annual Defense Report, for example, left no doubt of the emphasis to be placed on IT.[23]

From the 1970s until 30 August 1996 the Worldwide Military Command and Control System (WWMCCS) had networked the communication capa-bilities of the US military. This was necessarily dependent, in the first instance, on telegraph, telephone and radio, and, with the passage of time, on microwave and satellite links.

The transition from WWMCCS to GCCS was not all plain sailing. In a 1997 report, the DoD Office of Test and Evaluation concluded in their comments on the new GCCS-T system, to accommodate nuclear operations, that:

> After security testing, the Joint Staff granted GCCS(T) interim authority to operate at the Top Secret level. The Defense Special Weapons Agency

(DSWA) conducted parallel tests of their specialized capabilities and found no significant problems. However, testing also revealed the following anomalies with the non-DSWA capabilities, which are listed in descending order of operational impact: intermittent problems and delays logging on, occasional database error messages during stress testing, inability to send Focal Point (a commercial security application) messages from two sites, and serious formatting errors in printing at two of the five test sites. Because these problems were not debilitating and GCCS(T) performed much better than TS3, the Joint Staff proceeded with the transition and shut down TS3. In subsequent testing, GCCS(T) proved it can handle its designed workload, and all the former problems were corrected except the Focal Point requirement which was dropped by the Joint Staff.

Lessons Learned

Overall, the installation, initial configuration set up, and transition processes have proven very problematical in both GCCS and GCCS (T). These may be inherent difficulties with commercial architectures integrating products from many vendors and government sources. Improper configuration becomes a challenge and a principal cause of information warfare vulnerability. After testing, there are no widely effective tools and supporting policies for monitoring and enforcing configuration control.[24]

Because GCCS and its components are in continual development there is an explicit oversight of 'interoperability', the capacity of components to work as parts of the scheme.[25]

Since 1996 GCCS has gone through four versions; in 2004 GCCS 3.x was current and GCCS 4.x is anticipated in 2006.Q2. In turn, GCCS is to be followed by a new system, dubbed Joint Command and Control (JC2), with initial sections being ready in 2004.Q4.[26] While these convenient packaging names reflect major changes in architecture, pre-existing systems must be acknowledged, and changes are designed and implemented on an ongoing basis. DISA's chief technology officer said in April 2004 that DISA had released 27 updates of GCCS since 11 September 2001.[27]

Having committed to a new system, officials can speak more frankly about the shortcomings of GCCS and plans for the future. Any change in GCCS requires recompiling and retesting the entire system. JC2 will separate data transport, OS and Web services, applications and data, enabling independent updating.[28] Moreover, the DoD wants much greater bandwidth and a connected military. The planned network is called the Global Information Grid (GIG), the basis of 'net-centric warfare', with bandwidth extension: GIG-BE. The CEO of Lockheed Martin Corporation envisions 'a highly secure Internet in which military and intelligence activities are fused'.[29]

8.8   War experience: the Iraq War (2003–...)

Did the IT and GCCS features of the Iraq War (2003) play a significant – or even important – role in the US–UK battlefield success? Officials, and some observers, seem to think so.

On 9 April 2003 Vice-President Richard (Dick) Cheney told the American Society of News Editors that:

> Having been involved in planning and waging the Persian Gulf War in 1991 as Secretary of Defense, I think I can say with some authority that this campaign has displayed vastly improved capabilities, far better than we did a dozen years ago. In Desert Storm, only 20 percent of our air-to-ground fighters could guide a laser-guided bomb to target. Today, all of our air-to-ground fighters have that capability. In Desert Storm, it usually took up to two days for target planners to get a photo of a target, confirm its coordinates, plan the mission, and deliver it to the bomber crew. Now we have near real-time imaging of targets with photos and coordinates transmitted by e-mail to aircraft already in flight. In Desert Storm, battalion, brigade and division commanders had to rely on maps, grease pencils and radio reports to track the movements of our forces. Today our commanders have a real-time display of our own forces on their computer screens. In Desert Storm, we did not yet have the B-2. But that aircraft is now critical to our operations. And on a single bombing sortie, a B-2 can hit 16 separate targets, each with a 2,000-pound, precision-guided, satellite-based weapon.
>
> The superior technology we now possess is, perhaps, the most obvious difference between the Gulf War and the present conflict.[30]

Since Vice-President Cheney spoke, the Iraq War has become a war against urban guerrillas, who are heavily armed and, apparently, resilient. The USA's new surveillance and communications capabilities have been applied to this war, although whether they can provide a decisive edge remains to be seen. What we know, from anecdotal reportage of the wars in Afghanistan since September 11 and in Iraq, is that the US military still makes mistakes, and has not established conditions for its own security in either country. Early in the Iraq War a US unit operating with Kurds in the north was attacked by US aircraft, either because the US officer with the party calling in air support mistakenly gave his own coordinates rather than those of the intended target, or because of pilot error.[31] In Afghanistan, a US pilot attacked a group of soldiers on the ground; these turned out to be Canadian troops engaged in training exercises, four were killed and eight wounded.[32] In the fighting in Falluja in November 2004, a Marine squad, sitting on a dark rooftop, narrowly averted being struck by US aircraft, having been mistaken for 'insurgents'.[33] Nearly two years into the war the US forces had proven unable to secure the road leading to Baghdad Airport from the city, and required US personnel to go to the airport by helicopter.[34]

US officials do not dwell on such tragedies. Newly-appointed DoD head of intelligence Steven Cambone, a participant in the group which brought forth the September 2000 report of The Project for the New American Century, *Rebuilding America's Defenses*, told an interviewer that the melding of intelligence and 'operational arts' may have created a 'new mission area in its own right'.[35] On the day *The New York Times* reported the Cambone interview, it editorialized that 'an array of sophisticated spy planes and satellites, coupled with a computerised communications network, allowed the allies to see much more clearly than ever before what was happening on the battlefield'.[36]

8.9 Is GCCS sufficiently reliable for nuclear operations?

Are the computer and communications capabilities which support nuclear operations adequately stable and secure? In this section we will consider the issue of reliability. In the next we turn to security of SIPRNET, on which US nuclear operations and GCCS-T rely.

As a general rule, given sufficient time, computer systems will fail. At the time of writing, there is no evidence of GCCS failing, but we know that the WHAC systems were termed 'aged and failing' by the responsible official in 2003.[37] We do, however, know something about a major US intelligence system, failure of which could have had implications for military operations and nuclear deterrence. At the close of 1999, and again a month later, the National Security Agency (NSA) lost the ability to process data being downstreamed from satellites. How severe was this NSA breakdown?

On 29 January 2000 Reuters reported a NSA press release which said that: 'The National Security Agency headquarters suffered a serious computer problem at 7 p.m. (0000 GMT) on Monday, 24 January 2000.'[38] The system was restored on Thursday, after 72 hours.[39] NSA sought to insist that 'no significant intelligence information has been lost', while neglecting to assess the significance of *timeliness* in access to intelligence data:

> This problem, which was contained to the N.S.A. headquarters complex at Fort Meade, Md., did not affect intelligence operations, but did affect the processing of intelligence information...
>
> Contingency plans were immediately put into effect that called on other aspects of the N.S.A. system to assume some of the load. N.S.A. is confident that no significant intelligence information has been lost.[40]

The good news remains that testing is deeply embedded in understanding the use of GCCS, especially in respect of the Nuclear Planning and Execution

System. The mixed news, unsurprisingly, is that modification of software and hardware is a recurrent feature of the system. The bad news is that accuracy, reliability and usability may be found at any time to have become compromised, *even after testing and acceptance.*

8.10   Is SIPRNET sufficiently secure for nuclear operations?

The Secret Internet Protocol Router Network (SIPRNET) is a look-alike of our everyday Internet, but it has certain special characteristics. It is 'US only',[41] posing problems for working with coalition forces,[42] so severe that they have given rise to proposals that the 'US-only' stipulation be relaxed.[43] At least as important in the effort to ensure secure communications is SIPRNET's running on dedicated hardware: as designed, there is no way in which a user of the public-access Internet can access SIPRNET.

In order to be secure a network must meet several criteria:

- Secrecy of content: messages cannot be read.
- Integrity of content: messages cannot be altered.
- Assurance of delivery: messages cannot be destroyed.
- Assurance of authorship: sources and messages cannot be spoofed.
- Traffic can pass: no obstruction or significant delay.
- Immunity from traffic analysis: source, time, size, and addressees of messages are not accessible to unauthorized persons.
- Integrity of devices on the network: strangers cannot look at data without permission, and unwanted executables cannot be lodged.

Of course, these may not be 'messages' in the everyday sense, and may include file transfers, streaming data movement, webpage requests, and any other traffic.

Three different types of assurance are involved in the notion of a 'secure' network. Content and authorship are protected by encryption. Hardware redundancy and network architecture are chosen to enhance the likelihood that the system will work despite failures or disruption. A combination of wired (including fibre) and wireless hardware must move the traffic according to physical principles. The third issue, that of traffic analysis, does not exist if the network enjoys complete physical integrity, but if it can be 'tapped' at any point some data could become available to an intruder; commercial intrusion detection software is deployed.

We have described GCCS as a functional subnetwork running on the wide area network with the acronym SIPRNET, the Secret Internet Protocol Router Network, on which GCCS is heavily dependent.[44] SIPRNET embodies methods to confine the network to authorized users, and to enforce classification. Documents would be accessible only to those users who had the requisite clearance. As noted above, SIPRNET is a US-citizens-only network. A 'secure' network in this sense would have no direct interconnections with publicly-accessible telephone or data networks. SIPRNET is designed to be separate. On the other hand, in order to be useful there must be ways in which authorized personnel can access it.[45] Several measures are taken to confine access to authorized persons. DISA explains (emphasis added) that:

> The Defense Information System Network (DISN) has two separate Internet Protocol Router (IPR) Networks: the Secret Internet Protocol Router Network (SIPRNET) and the Unclassified but sensitive Internet Protocol Network (NIPRNET).
>
> The SIPRNET is a Wide Area Network (WAN) that is *separated both physically and logically from other networks*. Each access circuit and backbone trunk is encrypted to ensure integrity of information.
>
> SIPRNET uses several internetworking protocols to allow all types of traffic to traverse the network. These protocols include Internet Protocol (IP), Transmission Control Protocol (TCP), File Transfer Protocol (FTP), Telnet, Hypertext Transfer Protocol (HTTP) and Simple Mail Transfer Protocol (SMTP).
>
> Communications servers (Comm Servers) use secure data devices (STU IIIs) as the dial-up interface which are connected to DISA routers allowing STU III users to access the SIPRNET. STU IIIs restrict access to only authorized users by use of *an Access Control List (CAL), which is loaded into the STU III at the node site. Users must have a secret level SIPRNET user-key to be allowed connection* to the Comm Server. Further protection is added by use of the External Terminal Access Control Access Control System (XTACACS) which requires a login and password. For information on how to get Comm Server access, visit the Network Information Home Page.
>
> SIPRNET supports many of the important programs, such as the Defense Message System (DMS), the Global Command and Control System (GCCS) and the Global Combat Support System (GCSS).[46]

Bear in mind that GCCS-T imposes yet more demanding criteria for access, and, as it is said to be doubly encrypted, should be more strongly encrypted than ordinary SIPRNET.

Alternatively, once a person has an account authorizing access to SIPRNET, that person could acquire opportunities for mischief. You 'will be able to exchange email with any other user on the SIPRNET'.[47]

Certainly SIPRNET designers are aware of the possibility of transmitting executable code with e-mail, because such code could be used to compromise a user's machine. Several steps are taken to impede interference – in some cases employing the same software which you and I can buy from a local computer store; but e-mail attachments are specifically targeted:

> Services provided to fleet and shore customers include Secret Internet Protocol Network (SIPRNET) and Non-classified Internet Protocol Network (NIPRNET) connectivity. *Spectrum* software is used for network management. Firewall protection is accomplished by using *Gauntlet* software. A computer runs virus scans using *Norton Anti-Virus*. It is used for screening attachments to e-mail from customers outside the firewall destined to fleet customers inside the firewall.[48]

Are attachments sent from *within* the firewall tested? Apparently not.

This source includes an interview with D.B. Thomas, then the Information Systems Manager at the US Navy's Unified Atlantic Region Network Operations Centre. The interviewer introduces the issue of security:

> Certainly, military leaders have given consideration to the potential for security leaks. Since security is everyone's concern, I asked [Information Systems Manager D.B.] Thomas what he thinks of the risks versus the morale benefits.

> > Everything has risks. It's how we manage those risks that count. For instance, the CO has the capability to shut down and control e-mail from the ship. Additionally, we've installed *Norton Anti-Virus* software outside the firewall to scan attachments to incoming messages before they're passed to the fleet. However, the software scans for viruses, not for classification breaches and only the attachments are scanned not the basic email.

Thomas distinguishes SIPRNET, but seems to admit the possibility that SIPRNET could be entered (emphasis added):

> 'SIPRNET is a different story. That's cryptographically protected end-to-end. If someone hacks into SIPRNET, that's a major problem'.

> 'Throughout history, people have intentionally or unintentionally passed classified information. *There are people like Johnny Walker around. We've had them in the past; we've had them recently. We'll have them again.* But, the vast majority of our military people have the personal integrity and the training that keeps them from violating security rules', Thomas continues.

Could the NOC actually look at the messages' content, if they desired and had the man-hours? Thomas says, 'Yes we can, as well as check the logs and see where the messages are going. There's so much e-mail, and there aren't enough people to dedicate to gathering that data. We do spot checks for intended recipients, and we've found some amazing things'. The NOC is currently trying to locate funding so they can provide dial-in service outside the firewall. Then the NOC will be able to support additional shore-based commands in the local area.

The 'Johnny Walker' mentioned is John A. Walker, a US naval submariner and organizer of a spy ring of family and friends in the US Navy, which operated out of the naval port of Norfolk, Virginia. Walker collected secret material on naval cryptography and other subjects and passed it to the Soviet Union.[49]

A 1997 report appears to confirm that SIPRNET in the US Navy was vulnerable at that time. Despite designers' intentions, in a test of network security:

A United States Air Force officer managed to access the command and control system of a navy vessel at sea via the Internet and Siprnet, a military version of the Internet. (The loophole that enabled this has since been closed.)[50]

One concern for users of any military system is that sections of it might be destroyed in combat. Both to accommodate moving platforms, and to hedge against combat losses, the system incorporates Domain Name Server (DNS) redundancy:

The area of operations for the NOC is the Atlantic Region. However, they reach beyond that into the Mediterranean Sea, the Indian Ocean and the Persian Gulf.

When a ship transits out of the area covered by the NOC, the transition is nearly automatic. That's fairly new. 'With EURCEN up and running 100 percent, a large load has been lifted from the NOC. We no longer have to support all the ships from here. There's also a site in Bahrain. It's not a full-blown NOC but it provides network services for ships in the Gulf', says Thomas.

Take the DNS, for example. The NOC has secondary servers stationed throughout the world. If the primary DNS server goes down, another server would automatically key up and take over. In a network environment, the real backup lies in the secondary servers that are somewhere else . . .

Additionally, at the time of this report the Network Operations Centre (NOC) was 'upgrading the firewall' and 'preparing for the advent of the

NetRanger installation which monitors the networks for intrusions and hackers'.[51]

Clicking on a URL to a SIPRNET website from a NIPRNET machine yields the page 'Please note that you cannot hit a SIPRNET website from a NIPRNET machine'.[52] A commercial vendor offers, for $1,595, a five-day course in NIPRNET/SIPRNET Network Training for technicians who will be installing the systems, including issues such as 'strapping and installing cryptos' and 'crypto troubleshooting'.[53]

The nuclear SIOP was not brought into GCCS until a further Top Secret capacity was added. WWMCCS had been implemented as a Top Secret network with a high level of secrecy at TS3. On 30 June 1997, TS3 was turned off and GCCS-T, the Top Secret element of GCCS, was turned on.[54] What is it that distinguishes the 'top secret' level from the 'secret' level? Technically, GCCS-T implements double encryption – the implication being that ordinary GCCS does not. Users – whether military, government civilians, or contractor personnel – must have a corresponding security clearance. Where nuclear weapons are concerned, that will be a SIOP-ESI clearance in an appropriate category,[55] in turn requiring a recent-enough background check and certification that the user has a need for the access sought.

On the everyday Net users risk 'bad guy' efforts to compromise the *integrity* of computers and accessible data. There are well-known methods to hinder attempts to pry, download and exploit. Are these issues of the everyday world also issues for the world of 'secure computing' under government auspices? Computer scientist Eugene H. Spafford apparently thinks so, for he says of the US government that it uses:

> a monoculture computing system that has a compromised (and some would say, minimal) immune system. It is being used for weapons guidance, national defense, government, and communications. Most people use the same system on their personal and business computers... This system is the one on which we base our defense, our economy, and much of our scientific enterprise. It is increasingly under attack from malicious software.[56]

He then takes another implied swipe at Wintel systems by recalling that:

> The next generation of Navy aircraft carriers is going to have all weapons systems, propulsion, and command and control run by the very same system that you use at home to browse the Internet and play computer games.[57]

In summary, practical measures are being undertaken to assure a high degree of security for SIPRNET and GCCS; but vulnerabilities remain – vulnerabilities inherent in the need for personnel to use the system, and in inescapable technical facts.

8.11   Assessment

The review undertaken in this chapter highlights three problems for long-term, sustained nuclear weapons management. First, 'command and control' assumes that key personnel are available at crisis time, and that they have the information they need to make sensible decisions, but we know that this need not be the case. Second, 'command and control' requires a stable communications fabric, to provide information, enable discussion, and implement decisions; but we know that communications systems may fail. Third, controlling nuclear forces, and managing them against consequential error, requires that communications media be secure; and while we do not know of security failures in recent technologies, leaving aside old-fashioned spying, we can imagine a number of ways in which security could be breached. This does not lead to the conclusion that the efforts to design a stable and trusted military system to manage intelligence and communications have been unsound, or unserious, or ill-conceived, but rather that even the best of intentions cannot achieve perfect security.

'Cyberwar' implies cyber defence. Vulnerabilities should be understood as a constant concern of personnel assigned to maintain system integrity. As military IT systems grow more complex, so will the need to impose simplification and sustain secure practices. Hence, there is an Information Technology and Operations Centre at the US Military Academy, and West Point students engage in refereed competitions with their IT counterparts-in-training at the other US service academies. In Cyber Defense Exercises (CDX) in 2001 and 2002:

> An identical network of servers and workstations was set up at each school. During the first phase, teams of cadets and midshipmen at each site installed and configured an assortment of required services. The goal for each team during this phase was to configure the required service and the underlying operating systems in the most secure manner possible. In the second phase, an NSA-led penetration team attacked each site. This team, 'Red Team', conducted detailed reconnaissance and voluminous attacks over a five-day period. They maintained accurate records of any and all successful penetrations. A 'White Team' from CERT at Carnegie Mellon University refereed the exercise; they served as observers and controllers and, using an agreed upon scoring system, determined which school won.[58]

Planners concerned about the integrity of US nuclear command and control must also consider that the USA will not be alone in seeking military advantage through IT and a developed C4ISR capability. The current French draft programme governing military policy for the 2003–8 period stipulates that:

> Recent crises have confirmed the importance of technological superiority in the fields of intelligence, control and command, weapon precision and action at a distance in all its forms.

> Knowing this, certain new capacities must be acquired, in particular concerning intelligence, communications and command...[59]

What a traditional ally can do can also be carried out by states and groups which consider themselves to be adversaries. Can GCCS be imitated? While the conceptions that exploit IT are accessible, and the simple notion of secure broadband communications among widely-separated 'offices' is available to all, there are severe technological requirements at every point which render GCCS costly. A complete system must provide initial data acquisition, the effective processing of data initially in diverse formats and ascribed to distinct times and places, and the delivery to commanders and 'warfighters' in a timely and usable fashion. While it would be possible for high-tech societies to do those things that make up GCCS and implement a 'revolution in military affairs', the cost of doing *any* of them would be substantial (forcing a political consideration of the opportunity costs), and the cost of doing *many* of them prohibitive.

GCCS enables three aspects central to the RMA: (i) 'situation awareness', (ii) choosing more promising strategies and tactics, and (iii) prompt coordination. These correspond to the traditional requirements of war: knowing the battlefield, planning and decision, and execution. Actual warfare has always required adjustment to unexpected circumstance, enemy initiatives, and the success or failure of one's own measures. As designed, GCCS gives commanders more tools to fold performance and novel facts back upon the decision process, and to respond by choosing and carrying out new initiatives. On a more mundane level, it permits new and more precisely calibrated ways to meet logistic needs. In effect, GCCS conveys command of sequence, and so accomplishes a *revolution* by comparison to earlier methods. GCCS is a focused application of information technology to achieve far more than the organization of data available to individual service arms and programmes.

The smartest idea implemented in GCCS is *ongoing adaptability*. GCCS cannot be frozen, for the sensors, hardware, software, and man–machine interfaces through which military advantage is attained are the subjects of ongoing change. Moreover, as IT develops, creating both novel means of warfare and novel vulnerabilities, the *adaptability* of GCCS becomes increasingly central to US military doctrine. Transition to JC2 aims to make adaptation even easier. Now the DoD is moving to 'network-centric warfare', a GIG-BE network, implementation of the new Internet protocol IPv6, and enhanced solutions to the problem of rendering data usable.

Do these techniques promise the USA an *absolute* strategic and tactical superiority? The short answer, perhaps all that can be said confidently today, is that the USA can destroy with relative ease and impunity the military capabilities of any state which lacks nuclear weapons, but that its IT capabilities provide no advantage in converting battlefield victory to self-sustaining political gains. If the point of armed force in the twenty-first century is to achieve greater security and stability, which are essentially *political goals*, then neither IT nor talk of 'revolution' or 'networks' can carry that burden. The danger is that easy battlefield wins will deceive neo-conservative US strategists into believing that IT, the RMA and 'network-centric warfare' ensure an unchallengeable unilateral advantage, which can be exercised usefully outside a broad community of common political interest, outside a *collective security* structure. Being all-seeing and all-knowing, and endowed with unusual military capability, tempts empire. And empire fails.

Notes

1. Laura Hill, 'White House Communications Agency Transforms to Meet New Challenges', *Army Communicator*, Spring 2003, http://www.gordon.army.mil/AC/Spring/AC Spring 2003.pdf. Lieutenant Colonel Hill headed the Washington Area Communications Command.
2. Ibid.
3. March Laree Jacques, 'Transformation and Redesign at the White House Communications Agency', *Quality Management Journal*, 6(3) (1999).
4. Hill, 'White House Communications Agency'.
5. Ibid.
6. Ibid.
7. G.W. Bush, remarks to the 21st Century High Tech Forum, Washington, DC, 13 June 2002. http://www.whitehouse.gov/news/releases/2002/06/20020613–11.html.
8. US Department of Defense. Defense Information Systems Agency. FY 2005 Budget Estimate R-1 Exhibit, February 2004. Available at http://www.defenselink.mil/comptroller/defbudget/fy2005/budget_justification/pdfs/rdtande/DISA.pdf. The text further explains that 'Initial Operational Capability (IOC) has been tentatively scheduled for end of FY09 and is defined to be the deployment of the first CONUS AEHF satellite and the PNVC initiative implemented at the principal conferees' locations.' 'Another Defense component' is the National Security Agency, which will do most of the FY2005 work.
9. Richard V. Allen, 'The Day Reagan Was Shot', *The Atlantic Monthly*, April 2001. Subscriber access: http://www.theatlantic.com/issues/2001/04/allen.htm. All quotes in this section are from Allen's article.
10. National Commission on Terrorist Attacks Upon the United States. *The 9/11 Commission Report* (New York: W.W. Norton & Co., 2004). Hereafter cited as *Report*.

11. *Report*, p. 36.
12. Ibid.
13. Ibid. p. 37.
14. Dan Balz and Bob Woodward, 'America's Chaotic Road to War. Bush's Global Strategy Began to Take Shape in First Frantic Hours After Attack', *The Washington Post*, 27 January 2002.
15. Colin L. Powell, testimony before the US Senate Foreign Relations Committee, 25 October 2001. http://www.state.gov/secretary/rm/2001/index.cfm?docid=5751.
16. *Report*, p. 38.
17. Rear Admiral Nancy Brown, USN, Vice-Director of Command, Control, Communications and Computer Systems, Joint Staff. Testimony to the Subcommittee on Terrorism, Unconventional Threats and Capabilities, House Armed Service Committee, US House of Representatives, 3 April 2003. Available at, http://www.house.gov/hasc/openingstatementsandpressreleases/108thcongress/03–04–03 brown.html.
18. United States. Department of Defense. Defense Information Systems Agency, http://gccs.disa.mil/gccs/ (Revised 24 April 2002). This page continues with additional specifics about GCCS:

> The Global Command and Control System (GCCS) is the Department of Defense's computerized system of record for strategic command and control functions. With GCCS, joint commanders can coordinate widely dispersed units, receive accurate feedback, and execute more demanding, higher precision requirements in fast moving operations.

> Designating GCCS as the 'system of record' for strategic command and control implies that only data, which could include views, authorisations, and commands, which is held in GCCS itself, or which is linked to GCCS in such a way that the file changes only as changes are made within GCCS, is assured to be authentic and authoritative.

>> GCCS provides combatant commanders one predominant source for generating, receiving, sharing and using information securely. It provides surveillance and reconnaissance information and access to global intelligence sources as well as data on the precise location of dispersed friendly forces.

>> The GCCS enables warfighters to plan, execute and manage military operations. The system helps joint force commanders synchronize the actions of air, land, sea, space, and special operations forces. It has the flexibility to be used in a range of operations: from actual combat to humanitarian assistance. The Joint Operation Planning and Execution System (JOPES) has been transferred into this environment. The nation's military plans are kept and updated around the world using the JOPES system.

>> GCCS, built on modern information technology, is made up of database servers, applications servers, and clients. GCCS is a Command, Control, Communications, Computers, and Intelligence (C4I) system built on top of the Common Operating Environment (COE) and allows greater software flexibility, reliability, and interoperability with other computer systems. For example, commanders can establish their own homepages at the secret level and communicate securely through e-mail with counterparts around the

world. GCCS gives military personnel the same kind of 'plug and play' available with Windows-type operating systems on home computers.

GCCS includes support applications, such as the Automated Message Handling System (AMHS), and mission applications. The mission applications provide a unique set of capabilities and functionalities to assist the warrior by integrating data into a single tactical picture, replicating the data between GCCS sites, and displaying data in near real time on a single chart. Display filters can be used to construct different pictures in different map windows and overlays can be used to display air route data and airspace control.

19. For a more sophisticated description of the Defense Information Infrastructure (DII) Common Operating Environment (COE), see Carnegie Mellon Software Engineering Institute: http://www.sei.cmu.edu/activities/str/descriptions/diicoe_body.html; and Defense Information Systems Agency: http://diicoe.disa.mil/coe.
20. DISA, above. http://gccs.disa.mil/gccs/.
21. *1998 Annual Defense Report*, above, chapter 8. On 30 August 1996 WWMCCS was replaced by the Global Command and Control System (GCCS).
22. Recent overview documents outlining US defense policy are: Quadrennial Defense Review (QDR) (1997, 2001); Nuclear Posture Review (NPR) (irregular: 1994, 2001/2002); and Annual Defense Report (each January since 1995). Available at http://www.defenselink.mil/execsec/adr_intro. html. The 2001 QDR explicitly omits nuclear issues, deferring to the NPR then draft form. The QDR and NPR are issued in both unclassified and secret versions.
23. United States: Department of Defense, Secretary of Defense, *1998 Annual Defense Report*, chapter 8. This is spelt out in greater detail in the *Quadrennial Defense Review* [1997], as follows:

> The five principal components of our evolving C4ISR architecture for 2010 and beyond are:
>
> - A robust multi-sensor information grid providing dominant awareness of the battlespace to our commanders and forces;
> - Advanced battle-management capabilities that allow employment of our globally deployed forces faster and more flexibly than those of potential adversaries;
> - An information operations capability able to penetrate, manipulate, or deny an adversary's battlespace awareness or unimpeded use of his own forces;
> - A joint communications grid with adequate capacity, resilience, and network-management capabilities to support the above capabilities as well as the range of communications requirements among commanders and forces;
> - An information Defense system to protect our globally distributed communications and processing network from interference or exploitation by an adversary.

24. United States: Department of Defense. Office of Test and Development, *1997 Annual Report* Available at http://www.dote.osd.mil/reports/FY97/97tocmain.html. Accessed 24 July 2002. At archive.org.
25. Conveniently, the Joint Interoperability Test Command of the Defense Information Systems Agency maintains a frequently updated chart, which records the

state of certification. United States: Department of Defense. Defense Information Systems Agency. *GCCS Interoperability Status chart*. Available at, http://jitc.fhu. disa.mil/gccsiop/ and associated acronyms, http://jitc.fhu.disa.mil/gccsiop/ iop_table.htm.

26. Dawn C. Meyerriecks, speaking to the Systems and Software Technology Conference, Salt Lake City, Utah. *Global Computer News*, 21 April 2004. Available at http://www.gcn.com/vol1_no1/daily-updates/25644–1.html.

27. Ibid.

28. Ibid.

29. Quoted in Tim Weiner, 'Pentagon Envisioning a Costly Internet for War', *The New York Times*, 13 November 2004.

30. United States. The White House. 9 April 2003. Available at http://www.whitehouse. gov/news/releases/2003/04/20030409-4.html. (accessed 19 April 2003).

31. A BBC reporter was wounded, and a BBC translator killed; news of the attack, with graphic video footage, was broadcast around the world by BBC and rebroadcast, for example by NHK (Tokyo). On the attack, which killed at least 18, see Associated Press, 6 April 2003. Available at http://www.ctv.ca/servlet/ArticleNews/story/ CTVNews/20030406/iraq_friendly_fire_030406/World?s_name=&no_ads=.

32. CBC News, 19 April 2002, 'Canada Launches Inquiry into Afghanistan Bombing Deaths'. Available at http://www.cbc.ca/stories/2002/04/18/ cdndeaths020418.

33. Captain Read Omohundro 'heard on the radio that a group of about 15 insurgents had been identified somewhere close to his position, and that an air strike had been called in to kill them. Then something clicked in his mind, and he rushed to the radio and called off the air strike. The captain had been mistaken for an insurgent.' Dexter Filkins, *The New York Times*, 13 November 2004. Available at http://www.startribune.com/stories/484/5083665.html.

34. National Public Radio (Washington), 3 December 2004.

35. J.H. Cushman, Jr., and T. Shanker, 'War in Iraq Provides Model of New Way of Doing Battle', *The New York Times*, 10 April 2003.

36. *The New York Times*, editorial, 11 April 2003.

37. Hill, 'White House Communications Agency Transforms to Meet New Challenges'.

38. Reuters, 29 January 2000.

39. Ibid. and Associated Press, 31 January 2000.

40. National Security Agency statement, quoted in Associated Press, 31 January 2000.

41. United States: United States Marine Corps. Available at www.quantico.usmc.mil/ g6/ia/accessreq.doc. (Accessed 20 April 2003.)

42. Because SIPRNET is a 'United States-only' network, 'coalition operations' require special arrangements for information sharing. A 2001 US Army report explains in detail, a method for providing coalition partners with required information while protecting the classification integrity and 'US-only' status of SIPRNET. Major A. Jarvis, 'Information Sharing in a Coalition/Joint Headquarters', *News From the Front*, CALL, Ft. Leavenworth, Kansas, July–August 2001. Available at http://call. army.mil/products/nftf/julaug01/chap7.htm. Accessed 20 August 2003.

43. Speaking at the AFCEA Technet Asia-Pacific Conference in Honolulu, US officers called for greater access by coalition partners. One said [paraphrase] 'information sharing should include coalition partners' access to DOD's Secret Internet Protocol Router Network (SIPRNET), a move resisted by members of the intelligence

community'. Air Force Colonel Greg Brundidge, director of communications and information at Pacific Air Force headquarters, said 'the compelling operational requirement is the need to share, and we have to design systems that can do that.' *Federal Computer Week*, 15 November 2004. Available at http://www.fcw.com/fcw/articles/2004/1115/news-siprnet-11-15-04.asp.

44. 'GCCS employs client/server architecture using commercial software and hardware and open systems standards. Currently, GCCS integrates SUN, HP, and PC products and operating systems with ORACLE and SyBase distributed relational database support. This supporting architecture is now the 'Common Operating Environment (COE). GCCS communicates through local area networks (LANs) at CINC, Service, and Agency (C/S/A) sites worldwide, and which are tied together by the Secret Internet Protocol Router Network (SIPRNET) wide area network (WAN). GCCS can also be accessed from dial-in remote terminals. Altogether, the GCCS infrastructure is part of the Defense Information Infrastructure (DII).' United States. Department of Defense. Command and Control Research Program (CCRP), R.W. Anthony, Institute for Defense Analyses, *GCCS Evolution: Past, Present, and Future*, (n.d.: 1998). Available at http://www.dodccrp.org/Proceedings/DOCS/wcd00000/wcd000fd.htm. (Accessed 26 April 2003.) 1.2.1. The Command and Control Research Program resides within the Office of the Assistant Secretary of Defense (C3I).

45. A non-nuclear example is offered by ARAT, a programme to provide 'timely reprogramming of operational software in Aviation Survivability Equipment (ASE)'. This appears to enable rapid reconfiguration of defensive electronics in helicopters, aircraft, and ships. For this discussion, we are interested in the alternative *dial-up paths* which are built into the system:

> If you are a Warfighter and require access to the MDS, Tactics, Techniques and Procedures (TTP), and other related threat data files for your ASE, you will need to establish an account on the Multi-Service Electronic Warfare Data Distribution System (MSEWDDS) . . .
>
> For those who require additional intelligence data, or are looking for a more dependable and a more advantageous method to gain access to the MSEWDDS, you should turn to the Secret Internet Protocol Router Network (SIPRNET). The SIPRNET is just like the Internet, except encrypted (thus secure) at the SECRET Collateral level. Also just like the Internet, the SIPRNET has a World Wide Web (WWW) known as INTELink-S, which provides access to intelligence reports and information from a variety of National INtelligence Community agencies such as the National Imagery and Mapping Agency (NIMA), the Defense Intelligence Agency (DIA), and the National Ground Intelligence Center (NGIC). You can access SIPRNET through a direct drop at your location or via a direct dial-up SIPRNET account.
>
> To establish a direct SIPRNET drop at your location, you will need to co-ordinate directly with the Defense Information Systems Agency (DISA). A direct drop has its advantages, but it also is a costly and time intensive effort that doesn't lend practicality to mobile Warfighters. For this, the alternative is to obtain an ARAT dial-up SIPRNET account.

As mentioned above, the MSEWDDS site is accessible via SIPRNET/INTELink-S through a link on the ARAT SIPRNET website. This link allows users to access the MSEWDDS via INTELink-S SIPRNET web, provided they have an MSEWDDS account (userid and password). In addition, by attaining an ARAT SIPRNET account, you will also be able to exchange email with any other user on the SIPRNET.

United States. US. Army. [Internet] http://arat.iew.sed.Monmouth.army.mil/ARAT/ARAT_information/arat_services/arat_services.htm [Accessed 26 July 2002]. This source contains additional specific details about hardware and software required to effect a dial-up connection to MSEWDDS 'or the SIPRNET.' The hardware is identified explicitly. The required software, however, is everyday TCP/IP, PPP, and an ordinary Web browser.

46. United States: Department of Defense, Defense Information Systems Agency. [Internet] (Page last revised 14 May 2001). Available at http://www.pac.disa.mil/siprnet.html (Accessed 26 April 2003.).
47. Ibid., last sentence quoted in previous note.
48. United States: US Navy. E. Smith, Email and Internet services to the Fleet, a report on the US Navy's Unified Atlantic Region Network Operations Centre (UARNOC) (n.d., probably July 1998?). Available at http://www.chips.navy.mil/archives/98_jul/c_ews3.htm (Accessed 26 April 2003).
49. United States: Office of the National counterintelligence Executive. A biographic sketch of J.A. Walker. Available at http://www.ncix.gov/pubs/misc/screen_ backgrounds/spy_bios/john_walker_bio.html (Accessed 18 April 2003).
50. Lord Lyell and L. Ibrügger, 'Information Warfare and the Millennium Bomb', *AP 237 STC*, (97) 7, Draft General Report, 1 September 1997, p. 6, cited in NATO Parliamentary Assembly, Science and Technology Committee, Information Warfare and International Security, Draft General Report, 6 October 1999, Publication AS285STC-E. Available at (1999) http://www.naa.be/archivedpub/comrep/1999/as285stc-e.asp [Accessed 20 April 2003].
51. NetRanger is a CISCO Systems intrusion detection software product.
52. For example, try visiting http://jto.eustis.army.smil.mil.
53. CACI International, Arlington, Virginia. Available at http://www.caci.com/netcom/pdf/IP Network Description.pdf (Accessed 26 April 2003).
54. Anthony, *GCCS Evolution*, 1.2.2.
55. That SIOP-ESI categories exist is unclassified, and they are referred to as Category 1 and so forth, but description of the categories is classified.
56. E.H. Spafford, 'One View of Protecting the National Information Infrastructure', in A.H. Teich, S.D. Nelson and S.J. Lita (eds), *Science and Technology in a Vulnerable World*, July 2002, Supplement to AAAS Science and Technology Policy Yearbook 2003, Committee on Science, Engineering, and Public Policy, American Association for the Advancement of Science, pp. 41–42.
57. Ibid., p. 46.
58. Lt. Col. D. Ragsdale (presenter), 14 March 2003, 2003 Capital-Area Seminar on Information Assurance, UMBC Center for Information Security and Assurance, University of Maryland, Baltimore County [USA]. Colonel Ragsdale is director of the Information Technology and Operations Center. Available at http://cisa.umbc.edu/ (Accessed 18 April 2003).

59. France, Assemblée Nationale, *Projet de Loi relatif à la programmation militaire pour les années 2003 à 2008*, submitted to the Assemblée Nationale, 11 September 2002. Available at http://www.assemblee-nat.fr/12/projets/pl0187–1.asp (Accessed 20 April 2003).

'Les crises récentes ont confirmé l'intérêt de la supériorité technologique dans les domaines du renseignement, du contrôle et du commandement, de la précision des armes et de l'action à distance sous toutes ses formes'.

'Dès lors, certaines capacités nouvelles doivent être acquises, en particulier en matière de renseignement, de communications et de commandement...' Translation by the Ministère de la Défense: available at http://www.defense.gouv.fr/english/files/d140/index.htm.

9
Information Warfare and the Laws of War

Geoffrey Darnton

9.1 Introduction

Since the times of Grotius[1], an extensive system of international law in the public sphere has been developing. Increasing travel, trade, globalization, emigration, and cross-cultural marriage have also seen the emergence of a whole area of international law in the private sphere.

For several important reasons, this emergence of a body of international law is still in its infancy. The line of international law developed from Grotius has been predicated on the assumption of the sovereignty of nation-states and a system of public international law based, to some extent, on the idea of equality between sovereign states. In the area of public international law, there has been some progress towards the recognition of some key principles, but the major weakness still lies in the area of enforcement. Similarly, in the private international law arena, there is a complete absence of any international forum to handle matters of dispute, judgement and enforcement. The resolution of private disputes is still dependent on the jurisdiction of individual sovereign nation-states and their own legal systems, although in the arena of public international law nation-states have concluded a range of treaties, conventions and protocols governing various private law matters and mutual agreement between national jurisdictions.

Thus we are currently witnessing the simultaneous evolution of two aspects of international law of particular relevance to questions of information warfare (IW): the laws of war, and approaches to the resolution of conflict of laws in the private arena.

On the technological side, the most important development in recent years for both warfare and IW is the deployment and application of computer-based information and communications technology (cbICT). However, in the midst of the enthusiastic debate about ICT, we must not lose sight of the fact that what lies behind some information operations devices is far more extensive than cbICT, and may not even involve technology beyond the distribution of the information (for example, psychological operations, propaganda, and deceit).

A further ingredient in the relationship between IW and international law is the question of globalization. By this we do not refer to globalization in a purely commercial sense, although that may be where the focus of the debate is located. We are also witnessing globalization in the personal sense. Thus, an increasing number of people are migrating, and many parts of the globe are becoming increasingly multicultural. We are also witnessing many more cross-cultural links at a personal level in matters such as marriage, residence, domicile, pressure groups, interest groups, and personal contacts.

Globalization in both commercial and private spheres is underpinned by the application of ICT to both global information systems and personal, or individual, information systems (Darnton and Giacoletto, 1992).

Globalization is increasing tensions in the area of international law based on sovereign states, precisely because nation-state-based international law is usually of little help in matters of cross-border transactions in the business and private arenas. Increasingly, nation-based international law appears to lag far behind the needs of collective and individual entities operating in a multinational environment. More fundamentally, there are probably substantial cultural and political changes taking place. The potential depth of these changes should not be underestimated, and a useful location for debate is material concerning the emergence of a 'cyber culture' (Tofts et al., 2002). Thus, the greatest e-business today, by value, is foreign exchange, worth *billions* of dollars per day, which remains essentially unregulated internationally, and in the private sphere (and those who think of e-business as meaning the Internet and the World Wide Web, should bear in mind that the biggest e-businesses, by value, use communication channels other than the Internet and the World Wide Web).

Of particular concern to any debate between IW and international law is the relationship between military and civilian activities. There is an underlying presumption in international law that these should be kept as distinct as possible, particularly with reference to physical military assets and other physical assets used to sustain military capability or operations. However, the increasing inseparability of civilian and military technology and infrastructure is another emerging trend (Virilio and Lotringer, 1983; Levidow and Robins, 1989). To some people, the application of 'military discipline' in these areas may be seen as a positive step, but the world has experienced many instances of poor military thinking (Dixon, 1976) quite apart from the human limitations of regulation by military metaphor. This convergence between civilian and military infrastructure and a resulting convergence between military and economic interests may well prove to be a challenge that is fatal to the current regime of the laws of war. This could occur precisely because the fact that the development of any mechanism for the enforcement of the laws of war has been so fragile it offers a window of opportunity to nation-states to shatter the fragile system of international law.

9.2 Information Warfare (IW)

The possibility (likelihood?) that what we are witnessing at the present time is a profound merging of the military and civilian worlds, is reflected in much of the current literature about IW:

> The balance of the discussion seems to be going in the direction of IW being far wider than inter nation-state conduct, and it is not restricted to more traditional military contexts. What is being challenged today, is traditional concepts of both 'military' and 'warfare'. (Darnton and Rattanaphol, 2003).

Some writers have set out their view of the scope of IW. For example, IW may include:

- Software 'trap door' of public switching networks.
- Mass dialling attack.
- Logic bomb.
- Electronic takeover of radio or television.
- Video morphing.
- Alter medical formulas or information.
- Concerted e-mail attack.
- Divert fund or corrupt bank data.
- Steal and disclose personal information.
- Computer viruses or worms.
- Infoblockade.
- Disrupt a nation's command and control infrastructure (military).
- Manipulate or disrupt civilian infrastructure (stock or commodity exchanges, power, traffic control, navigation).

These have been summarized from Greenberg et al. (1997, pp. 3–6). Almost all of them are a long way from traditional military operations involving the immediate destruction of people or property, although they appear in a publication under the auspices of the US Department of Defense Command and Control Research Program (with the usual proviso that the views represented are solely those of the authors). Having said that, this list by Greenberg et al. is typical of what many writers are saying about the scope of IW (see, for example, Erbschloe, 2001).

Following this list, there are other kinds of operations that are not listed above, such as:

- 'Backdoors' in widely available hardware platforms.
- 'Backdoors' in widely available software.
- Sponsoring and linking to networks to enable mass surveillance.
- High industry concentration ratios for media.

Thus, the concept of 'warfare' has come full circle since the time of Grotius. In the middle of a confusing discussion of the etymology of the word 'war' by translators from Latin to English (Latin 'bellum' to English 'war'), the concept set out by Grotius (Grotius, 1682, book 1, chapter 1, section 2) was a broad concept along the lines I would construct as something like: 'what is war? [it is] a contention by force...a state of affairs...[it] comprises wars of every description...single combats are [not] excluded from this definition...and imply a difference between two persons'. Grotius' etymological analysis followed the Latin route from 'bellum'. Ironically, had Grotius followed an etymological analysis for 'war', he may well have developed a similarly broad concept: 'confusion, discord, strife...to bring into confusion or discord...in Old English the usual translation of *bellum* was...struggle, strife' (*OED*, 1989). The work of Grotius was taken by nation-states as a key foundation in the formation of the laws of war *as they apply to nation-states*. Hence, for some time, the concept of 'war' has been narrowed and assumed to be battlefield-oriented. If we need new international law to deal with non-state actors, Grotius provides an equally appropriate starting point. The current idea of IW has taken us right back to the original meanings of the term 'war'.

The 'civilianization' of warfare and military matters is not only taking place at a technological level. There is a technological civilianization prompted initially by:

i. Cost factors – the development of special military hardware is extremely expensive, and if general purpose material can be utilized, then costs can be reduced substantially.

ii. Economic ideology – the idea that competition and market forces produce the 'best' economic allocation and use of resources leads to an inexorable desire, on the part of those developing special military material, to obtain greater economic benefit by spinning off the developments into the civilian market.

Thus, we see some spectacular examples of initial military developments being spun off into the public domain. Examples include: personal computers and dramatic reductions in the size and weight of computers (the military need was for the miniaturization of computer components for space and missile programmes); global information systems (the military need is for global positioning and the development of precision weapons); the Internet (the military need is for a global communications system where military operations are global, and not just national).

There is a whole line of thinking about far deeper connections between the military and civilian worlds at cultural, political, economic and organizational levels:

> Computer-based models of war, work, and learning can promote military values, even when they apparently encourage the operator to 'think'. In all those ways, we are presently headed towards a military information society, which encompasses much more of our lives than we would like to acknowledge…In both spheres [army and industry] 'the psycho-physical apparatus of man is completely adjusted to the demands of the world, the tools, the machines – in short, to an individual "function"'. Discipline aspires towards the rational conditioning of performance, towards uniform and predictable behaviour…(Levidow and Robins, 1989, pp. 159–60)

This presents a frightening scenario of a subtle, unconscious and inexorable trend to reduce the primary role of each person to a 'function' within a larger system using military metaphors at least (and many military paradigms in practice) to legitimize the self-interested pursuit of power and economic advantage. Thus, we might have a fundamental cultural shift on our hands, which is merging military and human cultures – with the balance strongly in favour of the military (Virilio and Lotringer, 1983).

There is also an unfortunate side-effect of the western view of representative democracy. Leaders who engage in oppressive and unlawful (according to principles of international law) activities can claim to do so on behalf of the civilian voting population who elected, or had the opportunity to elect, the government. Therefore, other non-state actors can fall back on a line of reasoning that the general civilian population is culpable for electing the government concerned, and perhaps allowing apathy to prevail by surrendering decision making and power to those who wish to yield it in the name of the electorate. This is, of course, covered in part by Nuremberg Principles IV and VI, which, inter alia, warn citizens to be vigilant about the actions of their governments.

We should not, therefore, be surprised that we are now witnessing a major shift in military activities to include civilian areas and activities. This is particularly apparent in the case of non-state participants in 'warfare', and state participants who engage in psychological and information operations against a broader base than the enemies' military operations and personnel. The current 'war on terrorism' is against a vague ill-defined set of non-state participants who are alleged to make no distinction between civilian and military targets, and who indeed claim to do so in part because of the increasingly inseparability of military and civilian domains.

All of these developments pose some difficult challenges for international law in the coming years. While international law related to war was focused primarily on state military conduct, the tension between the actual state of international law and the ideal state led to a steady growth of widely accepted principles of the laws of war. However, the dramatic recent extensions in the scope of IW have resulted in a much greater list of issues to be addressed for the nature of international law to reach its 'ideal'. This difficult tension between a status quo based on existing nation-states and their sovereignty, and changes in the current international order, raises difficult problems for international law and the laws of war. IW and the emergence of the consequences of globalization only increase that tension. The increasing use of violence by non-state actors may well have the effect of entrenching the status quo. This tension has been explored by Falk (1966, p. 172):

> We live in a historical period when there is widespread awareness that survival, prosperity and security depend upon certain basic social and political transformations... [to] bring a warless world into being... without increasing the expansive capabilities of totalitarian societies... without increasing vulnerability to oppressive forms of government... one of the central obstacles on the path of transformation is the undesirability of freezing the fundamental political and social status quo...

9.3 Laws of war

The laws of war have never been a precise body of accepted legal thinking, and have different meanings for different people.

However, analysing a broad range of literature that appeals for some reason to the laws of war leads me to the proposition that the debate about the meaning of the term 'laws of war' takes place on three different dimensions, as set out in Table 9.1.

Table 9.1 Laws of war – dimensions of meaning

Dimension	Issues
Armed conflict vs unarmed conflict	• Rules concerning resort to (armed) conflict. • Conduct of (armed) conflict
State actors vs non-state actors	Who are the participants or combatants? Do legal rules apply independently of the formal status of the participants?
Material harm to people or property vs non-material harm to people or property	There are those who believe 'war' is only concerned with actions that involve physical harm to people or property.

The idea that there should be restraints on conduct in war is not new and dates back to ancient times (Roberts and Guelff, 1982, pp. 2–3). Following some medieval thinking (by Grotius in particular), the late nineteenth century saw the initial codification of important treaties and conventions. The original predecessor to the 1949 Geneva Conventions was the 1864 Geneva Convention (Pictet et al., 1952), drawn up following the initial formation of the Red Cross, to deal specifically with the issue of the wounded on the battlefield. Geneva is a small city, and clearly had a cluster of people focused on the issue of international law and the use of military force in international relations. Of course, Geneva itself, initially an independent republic, joined Switzerland following its occupation during the Napoleonic Wars. Geneva was also the home of the International Peace Bureau, founded in 1892, which worked to establish up the first Hague Convention in 1899, followed some years later by the 1907 Hague Convention (Darnton, 1989, pp. xiii–xiv). In 1910 the Bureau received the Nobel Peace Prize for its work. The International Peace Bureau has been continuing its work in recent years, helping to galvanize efforts to bring the question of the legality of nuclear weapons before the International Court of Justice in the Hague (Mothersson, 1992).

Of critical importance to the question of international law applied to IW is the extent to which existing international law, some of which is more than 100 years old, can apply to a new situation that was simply not envisaged at the time that the various treaties, conventions and protocols were agreed. The possibility of this situation arising in the future was in the mind of the drafters from the start, presumably because of an awareness of the rapidity of technological change, and a recognition of human ingenuity in devising new ways of harming others. The clearest unambiguous adoption of such a principle to apply to the future lies in what has become known as the 'de Martens clause' from the preamble to the 1907 Hague Convention IV Respecting the Laws and Customs of War on Land:

> Until a more complete code of the laws of war has been issued, the high contracting Parties deem it expedient to declare that, in cases not included in the Regulations adopted by them, the inhabitants and belligerents remain under the protection and the rule of the principles of the law of nations, as they result from the usages established among civilized people, from the laws of humanity, and the dictates of the public conscience. (This has been reproduced in many sources, but can be found in Darnton, 1989, pp. 1–7)

The de Martens clause should clarify an essential point in the application of international law. Some lawyers have questioned the applicability of this clause as it is 'only' in the preamble, but this has been considered, and the International Court of Justice (ICJ) has made full use of the preamble in case law (Singh and McWhinney, 1989, p. 47). There are ultra-statists who take the concept of state sovereignty to mean that unless something is explicitly prohibited in international law, it is permitted. An illustration of the ultra-statist view is provided by Greenberg et al. (1997, p. 17) who state:

> Perhaps because of the newness of much of the technology involved, no provision of international law explicitly prohibits what we now know as information warfare. This absence of prohibitions is significant because, as a crudely general rule, that which international law does not prohibit, it permits. But the absence is not dispositive, because even where international law does not purport to address particular weapons or technologies, its general principles may apply to the use of those weapons and technologies.

The de Martens clause has been considered subsequently by the ICJ. The unanimous opinion of the Court concerning the applicability of humanitarian law has dealt a fatal blow to the ultra-statists: 'I see the limitations laid down in paragraph 1(c) [of the Opinion] as laying that argument to rest' (ICJ, 1996, Dissenting opinion of Judge Weeramantry, Preliminary Observations (c) (iv)).

The de Martens clause was effectively re-stated in a slightly different form in the Protocol Additional to the Geneva Conventions of 12 August 1949, and relating to the Protection of Victims of International Armed Conflicts (Protocol 1) of 1977, particularly by Article 1 (2):

> In cases not covered by this Protocol or by other international agreements, civilians and combatants remain under the protection and authority of the principles of international law derived from established custom, from the principles of humanity and from the dictates of public conscience.

This question of the applicability of international law to new forms of warfare has been alive for many years with respect to nuclear weapons. The nuclear weapons states have justified the possession and possible use of nuclear weapons on the ultra-statist position that nuclear weapons per se have not been declared – or rather agreed to be – unlawful in international law (presumably in part because the nuclear states are hardly likely to agree to any such classification). The question was sent by the United Nations (UN) General Assembly to the International Court of Justice (ICJ) for an Advisory Opinion. The Advisory Opinion on the 'LEGALITY OF THE THREAT OR USE OF NUCLEAR WEAPONS' was delivered dated 8 July 1996 (ICJ, 1996). It contains two important paragraphs illustrating the current applicability of the de Martens clause:

> . . .
> 85. Turning now to the applicability of the principles and rules of humanitarian law to a possible threat or use of nuclear weapons, the

Court notes that doubts in this respect have sometimes been voiced on the ground that these principles and rules had evolved prior to the invention of nuclear weapons and that the Conferences of Geneva of 1949 and 1974–1977 which respectively adopted the four Geneva Conventions of 1949 and the two Additional Protocols thereto did not deal with nuclear weapons specifically. Such views, however, are only held by a small minority. In the view of the vast majority of States as well as writers there can be no doubt as to the applicability of humanitarian law to nuclear weapons.

. . .

87. Finally, the Court points to the Martens Clause, whose continuing existence and applicability is not to be doubted, as an affirmation that the principles and rules of humanitarian law apply to nuclear weapons.

. . .

Thus, as a useful starting point, there are good grounds here for asserting that if IW really is a new form of warfare, then customary international law, in the form of what is usually termed humanitarian law, is applicable. There are three fundamental aspects of this stated in the de Martens clause:

- the usages established among civilized people,
- from the laws of humanity, and
- the dictates of the public conscience

Therefore, contrary to the position taken by Greenberg et al., the applicability of the de Martens clause and all other elements of humanitarian law, including the dictates of the public conscience, is not in doubt, even though understandably none of the current instruments of international law render IW to be unlawful per se.

There is in fact a legal requirement that this question of the applicability of international law be considered by the signatories to the 1977 Geneva Protocol I, by virtue of Article 36:

Article 36. – New weapons. In the study, development, acquisition or adoption of a new weapon, means or method of warfare, a High Contracting Party is under an obligation to determine whether its employment would, in some or all circumstances, be prohibited by this Protocol or by any other rule of international law applicable to the High Contracting Party.[2]

The 'instinctive' application of international law to IW represented initially by so many writers who present a wide concept of IW is reasonable, and can be taken as initial expressions of 'usages among civilized peoples' and 'dictates of the public conscience'. There have been few instances of serious attempts to measure the 'dictates of the public conscience' for the purpose of legal relevance. One such example is the London Nuclear Warfare Tribunal in 1985, which took formal evidence from several witnesses on this specific point, with those witnesses being cross-examined by professional legal counsel. Several sources of information about the public conscience were considered, and in that tribunal were derived from representatives of religious groups, representatives of protest and mass movement groups, notions of individual criminal responsibility, individual protesters involved with controversial topics, and further commentary in various treaties. As another example, the ICJ received substantial input from the public (Grief, 1997, p. 681ff at 684, note 24) while it was forming its opinion:

> ... nearly two million signatures have been actually received by the Court from various organizations and individuals from around 25 countries. In addition, there have been other shipments of signatures so voluminous that the Court could not physically receive them and they have been lodged in various other depositories. If these are also taken into account, the total number of signatures has been estimated by the Court's Archivist at over three million (ICJ, 1996, dissenting judgement by Judge Weeramantry, I(1)).

It is useful at this point to summarize the key customary norms of international law:

Principle of Discrimination – to be lawful, weapons and tactics must discriminate clearly between military and non-military targets, and be confined in their application to military targets. Indiscriminate warfare is *per se* illegal, although indirect damage to civilians and civilian targets is not necessarily so. **Principle of Proportionality** – to be lawful, weapons and tactics must be proportional to their military objective. Disproportionate weaponry and tactics are excessive, and as such, illegal. **Principle of Lawfulness** – to be lawful, weapons and tactics must not violate any treaty rule of international law binding as between the parties. **Principle of Necessity** – to be lawful, weapons and tactics involving the use of force must be reasonably necessary to the attainment of their military objective. No superfluous or excessive application of force is lawful, even if the damage done is confined to the environment. **Principle of Humanity** – to be lawful, no weapon or tactic can be relied upon that causes unnecessary suffering to its victims, whether by way of prolonged or painful death, or in a form that is calculated to cause severe terror or fright. For this reason, weapons and tactics that spread poison, disease, or do genetic damage are generally illegal *per se*, as being weapons with effects not confined in the place and time of damage to the battlefield. Such a prohibition, under contemporary circumstances, extends to ecological disruption in any form. **Principle of Neutrality** – to be lawful, no weapon or tactic can be relied upon that seems likely to do harm to human beings, property, or the natural environment in neutral countries. A country is neutral if its government declares itself to be so and if it pursues a policy of impartiality in relation to armed conflict, including the avoidance of any kind of alliance relationship. (Darnton, 1989, pp. 1–5 to 1–6).

This provides a useful framework within which to judge the relevance of international law to a broad view of IW. Traditional military warfare is ideally confined to damage inflicted on the battlefield at the time and place of battle, and distinct from civilian areas. In addition, suffering should be kept to a minimum – with access and help available to the wounded. Any departure from that scenario raises the possibility of some violation of the principles set out above.

On this kind of analysis it is clear that IW poses serious challenges to the principles of international law. From the kind of list enumerated above from Greenberg et al., it can be seen that an immediate difference between traditional warfare and IW, is that, apart from some specific information operations, what IW envisages are operations not primarily focussed on the immediate time and place of a battlefield. IW operations are calculated to have more delayed effects, such as disruption of command and control, disruption of infrastructure, disruption of logistics and supply chains, disinformation, and the alteration of beliefs, attitudes, and behaviour. It is also abundantly clear that many possible information operations are directed at infrastructure. Can this kind of activity be lethal, or cause serious damage to people or property? In many instances the effect is not as immediate as battle, but may be as lethal in a longer time period. As far as I am aware there have been no formal legal considerations of this question. Therefore, it is only possible to speculate based on anecdote.

In 1990 in the USA a software error shut down a significant number of long-distance telephone lines for several hours (Lee, 1991). This resulted in serious disruption to all kinds of services and has been estimated to have cost hundreds of millions of dollars of disruption. Some people even died as a consequence of communications failures. This problem is usually attributed to the introduction of a small un-tested software change. However, another more significant change was that in order to reduce costs (and therefore increase profits) newer telecommunications networks were being designed with lower levels of redundancy in the systems, thus increasing the level of vulnerability. A common feature of recent developments in the application of ICT has been the creation of longer and 'thinner' supply chains and communications lines, which has increased the level of vulnerability. Therefore IW becomes a more serious issue for an economy that becomes more 'modern' through the deployment of ICT. This single incident over telephone circuits highlighted some potentially serious consequences of IW. There are many others apparent from our experiences with computer and software errors and failures (Lee, 1991; Neumann, 1995).

When supply lines for civilian facilities such as supermarkets and fuel are long and thin, and people have lifestyles based on low buffer stocks, disruption to communications could lead to serious civil unrest in a comparatively short period of time. This kind of deliberate action is clearly a violation of several of the principles of humanitarian law.

Table 9.2 sets out some ways in which the results of IW could be in violation of humanitarian principles established in international law in relation to war.

A greater challenge for international law is to bring non-state actors and participants within a legal framework that applies the principles of humanitarian law. It is disingenuous for an individual or group, not being a state-actor, to claim that general principles of morality or the dictates of the public conscience do not apply to their actions, because they are not, per se, state actors. Of course, the counterpart to that is that state actors need to be much more accountable for their violent and oppressive conduct towards others, because it is that hypocrisy that is often used by non-state actors to justify behaviour that is normally considered barbaric by the usual dictates of public conscience. It is inappropriate to classify certain acts as 'terrorism' when the acts are carried out by a non-state actor, whereas the same acts are not considered to be 'terrorism' if carried out by a sovereign state actor. There is much credibility in the complaint that today state terrorism is more extensive than non-state terrorism.

Turning to private international law, there is a whole vacuum in terms of contract formation, jurisdiction and enforcement for the increasing number of individuals and corporations who operate internationally for a variety of reasons, including forming cyber tribes or cyber communities. That vacuum

Table 9.2   Humanitarian law principles and Information Warfare

Principle	*Possible violations by Information Warfare*
Discrimination	Many scenarios posited for IW are in direct violation of this because they are targeted precisely at civilians and civilian infrastructure.
Proportionality	Many forms of IW do not have any clear military objective, therefore they cannot be proportional to anything.
Lawfulness	Many proposed forms of IW do not violate many laws of war per se, but do violate a whole range of international law concerning things like the disruption or spoofing of communications.
Necessity	Many forms of IW are not directed at military objectives, therefore they are not 'necessary'.
Humanity	Many projected forms of IW are not projected at a battlefield, and resulting disruption could cause widespread terror, fright, and disruption of essential humanitarian supplies.
Neutrality	Given the ubiquitous nature of information and IW, it may be difficult to ensure that IW operations do not make any use of facilities in neutral countries or cause collateral risks and detriment in neutral countries.

is being filled in part by commercial non-state actors such as credit card companies and banks. State actors are increasingly irrelevant.

9.4 Key issues

This chapter has raised some fundamental issues about the relationship between international law and IW. There is no doubt that current principles and applicability challenge international law as it exists at present.

There are some clear improvements needed to develop an international legal framework to cover IW:

- Need for state actors to take initial responsibility to ensure that they themselves operate according to the 'rule of the principles of the law of nations, as they result from the usages established among civilised people, from the laws of humanity, and the dictates of the public conscience'.
- Bring non-state participants within the same principles of the laws of humanity.
- Strengthen international law to cover many of the new anticipated IW operations.
- Introduce publicly available international institutions (publicly available means available independently of the consent of sovereign states) to enforce relevant international law (at the moment it seems that many aspects of information war are only checked by national laws or private contractual mechanisms).
- The emergence of IW in such a broad form will mean that the intersection between international law and domestic law will become even greater. A useful model for doing that, also from the consideration of the legality of nuclear weapons and domestic involvement, is the development of a line of domestic cases within the UK, drawing on a combination of international and domestic law (Manson, 1995).

However, notwithstanding that there are serious weaknesses in current international law to deal with the threats of IW, the analysis set out above shows that there is an established body of law that does have applicability to IW. The most worrying of the gaps is caused because of the civilianization of many kinds of IW operations, and the potentially serious large-scale humanitarian problems that could be caused by IW.

Notes

1. Hugo Grotius (1583–1645) is considered by many to be the starting point for modern thinking about the laws of war. His major work, *De Jure Belli ac Pacis*

[On the Law of War and Peace], first appeared in Latin in Paris, 1625. Availability in English has helped to propagate his ideas much more widely (see Grotius, 1682).

2. The ideas in this chapter were discussed at the Second European Conference on Information Warfare, Reading, July 2003. During discussion, it emerged that at least two countries (the UK and the USA) have probably carried out such Article 36 assessments, but that the results are secret!

References

Darnton, G. (ed.), *The Bomb and the Law: London Nuclear Warfare Tribunal Evidence, Commentary and Judgment* (Stockholm: Alva and Gunnar Myrdal Foundation, 1989).

Darnton, G. and S. Giacoletto, *Information in the Enterprise: It's More Than Technology* (Burlington, MA: Digital Press, 1992).

Darnton, G. and J. Rattanaphol, *RMA Applied to Thailand: European Conference on Information Warfare*, 2nd edition (Reading, England: MCIL, June 2003).

Dixon, N.F. *On the Psychology of Military Incompetence* (London: Jonathan Cape, 1976).

Erbschloe, M. *Information Warfare: How to Survive Cyber Attacks* (New York: Osborne/McGraw-Hill, 2001).

Falk, R.A. 'Historical Tendencies, Modernizing and Revolutionary Nations, and the International Legal Order', in R.A. Falk and S.H. Mendlovitz (eds), *The Strategy of World Order* (New York: New York Law Fund, 1966), vol. 2, pp. 172–88.

Greenberg, L.T., S.E. Goodman, and K.J Soo Hoo, *Information Warfare and International Law* (Washington, DC: National Defense University Press, 1997).

Grief, N. 'Legality of the Threat or Use of Nuclear Weapons', *International and Comparative Law Quarterly*, 46 (1997), 681–8.

Grotius, H. *The Most Excellent Hugo Grotius. His Three Books Treating of the Rights of War and Peace* (London: Thomas Basset, 1682). (Text readily available on line. See, for example, http://www.ecn.bris.ac.uk/het/grotius/Law2.pdf.)

ICJ, *Advisory Opinion on the Legality of the Threat or Use of Nuclear Weapons* (Hague: International Court of Justice, 1996).

Lee, L. *The Day the Phones Stopped: the Computer Crisis – the What and Why of It, and How We Can Beat It* (New York: Donald I. Fine, Inc., 1991).

Levidow, L. and K. Robbins, 'Towards a Military Information Society?', in L. Levidow and K. Robbins (eds), *Cyborg Worlds: the Military Information Society* (London: Free Association Books, 1989).

Manson, R. *The Pax Legalis Papers: Nuclear Conspiracy and the Law* (Oxford: Jon Carpenter, 1995).

Mothersson, K. *From Hiroshima to the Hague: a Guide to the World Court Project* (Geneva: International Peace Bureau, 1992).

Neumann, P. *Computer-related Risks* (London: Addison-Wesley Publishing Company, 1995).

OED, *Oxford English Dictionary*, 2nd edn (Oxford: Oxford University Press, 1989).

Pictet, J.S., F. Siordet, C. Pilloud, J.P. Schoenholzer, R.J. Wilhelm and O.H. Uhler (eds), *Geneva Convention for the Amelioration of the Condition of the Wounded and Sick in Armed Forces in the Field: Commentary* (Geneva: International Committee of the Red Cross, 1952). (This set is what is usually supplied on asking for a copy of the 'Geneva Conventions'.)

Roberts, A. and R. Guelff (eds), *Documents on the Laws of War* (Oxford: Clarendon Press, 1982).

Singh, N. and E. McWhinney, *Nuclear Weapons and Contemporary International Law*, 2nd revised edition (Dordrecht: Martinus Nijhoff Publishers, 1989).

Tofts, D., A. Jonson and A. Cavallaro (eds), *Prefiguring Cyberculture: an Intellectual History* (Cambridge, MA: MIT Press, 2002).

Virilio, P. and S. Lotringer, *Pure War* (New York: Semiotext(e), 1983).

Part III
Country Perspectives

10
RMA: The Russian Way

Fanourios Pantelogiannis

10.1 Historical overview

By the mid-1970s, especially after America turned more isolationist following the debacle of the Vietnam War, the former Soviet Army had become the strongest military machine in the world. However, from that time on, as western countries moved ahead into an era of 'smart', super-accurate weapons, the Soviet military became relatively weaker. By the second half of the 1980s, it was clear that the Soviet military had lost its air of superiority; for example, in Afghanistan, it could not find an effective countermeasure to the American Stinger anti-aircraft missiles.

Within this framework, the 'new' military force of the Russian Federation was established in 1992 to replace the Unified Armed Forces of the Commonwealth of Independent States (CIS). The loss of some border areas of the former Soviet Union led to loss of a large part of the most modern military hardware, infrastructure and parts of the Soviet Union's air defence system.

Currently, military depots across Russia still hold huge arsenals, but the number of up-to-date weapons is diminishing and troops receive little training. Due to fuel shortages, for example, Russian pilots fly an average of 25 hours a year, compared to the West's minimum of 200 hours. The army has only around 5–7 per cent of equipment which can be classified as new (less than ten years old), with no more than around one-quarter of the equipment and military hardware being combat ready.[1] Modern types comprise only about 40 per cent of the tanks and infantry fighting vehicles, 30 per cent of the Surface-to-Air Missile (SAM) and artillery systems and 2 per cent of the helicopters. The Russian pilots, for example, still fly aged Mi-8 and Mi-24 helicopters.

Thus, the Russian military remained, to a great degree, a product of the first half of the twentieth century. However, the failure of Russia to successfully cope with modern technical realities did not appear to prevent thinking about future war, and/or Information Warfare (IW). From the 1960s, Soviet thinkers were among the first to postulate and analyse the implications of what they called the Revolution in Military Affairs (RMA) or the 'scientific-technical revolution' (STR) or, in Russian, the NTR.[2] Indeed, Soviet writers coined the term 'Revolution in Military Affairs' and developed the concept before US writers and officers appropriated it.[3] In the Russian definition, new weapons could include biological or psychotropic weapons, or simply innovative forms of information and other technologies with the intention of destabilizing a society from within. Indeed, today's Russian writers on this subject are as visionary as were their predecessors in the 1970s and the 1980s.

Nevertheless, the Russian RMA did not remain a purely theoretical canvassing. Recent wars, in particular the conflicts in Kosovo and Afghanistan and the operation 'Iraq Freedom', demonstrated that maintaining vast armies is of little use. The RMA, being intended to prepare military forces for highly demanding contingencies, manifested itself in rapid and decisive force employment, and reduced levels of vulnerability. 'Our military brass was saying... that the American bombing campaign in Afghanistan would produce no results. The sudden unravelling of the Taliban shocked them,' says Alexander Goltz, a military expert with the weekly magazine *Ezhenedelny Journal*. Anatoly Medetsky adds: 'The ability of the US to wage remote-control warfare, to destroy its enemies without a costly effort of occupying territory, has been very eye-opening for our military leaders'.[4]

Russian officials thus understood that in Russia's most-likely future wars, the terrain will probably not support large armoured formations and that therefore C4ISR (Command, Control, Communications, Computers, Intelligence, Surveillance and Reconnaissance) and Electronic Warfare (EW) systems should govern the allocation of scarce defence resources. The era of large-scale wars and, consequently, massive armies was about to pass for Russia.

Adopting this strategy clearly required several major changes. And after years of 'virtual' military reform, experts say that the Kremlin is now serious in its intentions. Defence started being steadily replaced by military affairs and the focus shifted from the quantity of personnel and arms to their quality. In exactly this framework, this chapter proposes to offer an overview of the current situation of the Russian RMA, its driving factors and its regional and international consequences. Evaluating the current situation as well as the historical process of the subject will help us to understand the paradox of Russian General Staff thinking, as well as make an assessment of possible future trends.

Reforming the Defense Ministry – that is to improve the existing army – is impossible. Not that it is bad to itself. The main reason is that it was created to pursue different objectives. It is an army from the last century, mobilizing all national resources. What it is not is a high-tech, modern army, which serves those resources and 'covers' them. There is only one way out of the situation: change over to a strictly professional army. One created on a new foundation. With new goals. For the new century. Will Russia have the sufficient political courage and financial funds to make the breakthrough? This is the gist of the matter.[5]

10.2 The current Russian RMA and its international consequences

The question that might be posed, of course, is why Russian views on the RMA should be studied. Firstly, over the last few years we have been witnessing a dramatic metamorphosis in warfare. As Engels pointed out, the development of military capacity depends 'on material (economic) circumstances, on human resources, and weapons, which, in turn, depend on the quality and size of a nation's population and on its technology'.[6] Without considering these factors no light can be shed on the organization and personnel of a nation's armed forces or on its military policy and strategy. And the question of what specific aspects of the RMA a nation such as Russia might obtain, could have a profound influence on future global war and peace trends. As the Russians say, in the near future, the primary competitors in the RMA will be the USA, Russia, Japan and possibly China, probably in that order.[7]

Many of the earliest calls for change came from the Soviet military theorists examining the impact of the so-called 'Military-Technical Revolution'. Officers such as Marshal Orgakov of the Soviet Union drew attention to the new weapons technologies which would be available at the turn of the century and argued that in combination they would lead to changes in warfare that would be as far-reaching as the introduction of battlefield nuclear weapons had been. In November 1997, Andrea Nikolayev, a State Duma deputy and army general, wrote in *The Russia Journal*: 'How can we improve our military potential without a clear picture of what modern war is all about?' A response to this question was given two years later by Russia's 1999 Draft Military Doctrine:

> Regional war...will be characterized by:...warfare in all spheres; coalition operations; mass use of PGMs, electronic and other new forms of combat; attacks throughout the territory of the opposing sides... An Independent S&T, technological and productive base must be developed to meet the military's needs, especially for a new generation of armaments. Priorities in this area are: qualitatively upgrading strategic arms; developing C3I and fire control systems; strategic warning; electronic warfare; mobile, precision non-nuclear weapons and their information support; reducing and standardizing the numbers of different weapons and equipment.

A few years later, Captain Vladimir Barinov, senior assistant to the Military Representative (MILREP) of Russia to North Atlantic Treaty Organization (NATO), stated in front of the committee:

> The following are features of the development of tactical weapons for Russian general-purpose forces:...creation of high accuracy weapon systems,... creation of advanced individual soldier's equipment with elements from new-generation combat and auxiliary systems;...development of an advanced front-line fighter with extended capability to engage ground targets;...The principal trends in general-purpose naval force development in the next ten years are: creation of new-generation multi-purpose submarines; design and quantity production of dual-purpose surface ships with high-accuracy assault and anti-submarine weapons;... creation of new ship borne and land-based aircraft; further development of C3 equipment.

Unfortunately, the exact numbers, strength and combat ability of the Russian army remain a military secret and we can refer to them only by assessments and statistics. However, it is commonly accepted that the army is in need of urgent reform. To respond to security challenges in Central Asia and elsewhere, for example, Russia needs a highly mobile, high-tech military force, with the ability to fight in a broad variety of terrains and to deploy at short notice. Creating that kind of force is exactly what President Putin committed to do in November 2001, to the great vexation of his generals.

However, for the time being, the situation of the Russian Armed Forces is far from ideal. At present, arms procurement accounts for 6 per cent of total defence expenditure, compared to a minimum of 20 per cent in NATO member states. Since the collapse of the Soviet Union, the armed forces have not purchased any military transport planes, apart from a few IL-76s. There has also been no mass production of new technology such as the KA-52 helicopters. The majority of the money the military receives goes to maintenance and modification of old arms and technology, with a small part of the budget going to the development of prototypes. 'If such a situation persists for another five to six years, our army will turn into a museum, not into armed forces capable on defending the country.'[8]

The Russian space forces have half the number of satellites of the US Space Command,[9] and their number 'is decreasing faster than new ones are being launched'.[10] Furthermore, 70 per cent of the Russian satellites have already expended their service life and therefore it is not certain that they can fulfil their original tasks such as strategic intelligence, early warning of missile attacks and communication. This is, of course, an expected result of the fact that the Soviet space programme had reached its apogee during the 1960s and the 1970s.

In addition, most of Russia's heavy missiles will be withdrawn from service by 2008, having long since passed their guaranteed service life. SS-19s (Stiletto) will rapidly reach the end of their operational life after 2007,[11] the SS-24 (Scalpel) will be phased out by 2007 and few, if any, SS-18s (Satan) and SS-25s (Sickle) will remain by the end of 2010. The number of operational Sea-Launched Ballistic Missiles (SLBMs) may equally drop over the next decade as nuclear-fuelled Russian Ballistic Missile Submarines (SSBNs) reach the end of their service lives.[12] In total, 60 per cent of Russian Intercontinental Ballistic Missiles (ICBMs) are past their guaranteed life, half of the operational SSBNs (and 75 per cent of their missiles) and most of the ICBM warheads require replacement by this year (2005) at the latest. It is unlikely that Russia will be able to build new missiles of this calibre. This leads us to the most modern SS-28s (Topol-M).[13] It had been hoped that Moscow would be able to produce 30 new SS-28s each year, but the reality is no more than ten a year.[14] The available maintenance infrastructure is also in a parlous state, starved as it is of equipment and parts required for making repairs.

In the air force, up to 2005, there had been no realistic prospect of receiving funding for new aircraft.[15] As a result, faith was being placed in upgrading existing models. The basic MiG-29 fighters are therefore being converted to MiG-29UBT and MiG-29Smt standards in order to increase the aircraft's air-to-ground capabilities. Similarly, although later on, numbers of Su-27 will be upgraded to the Su-27IB version to form the basis of strike and reconnaissance aviation. After those projects, Russian officials are thinking of improving the old Su-24 and Su-25 as well as the MiG-31. These projects are inevitably nothing more than stop-gap solutions. According to Russian officials, real solutions have already been developed and only require money for their production to begin. In accordance with the principle of reducing to a minimum the types of aircraft fielded, 'There will be one multi-role aircraft, the Su-34. It has been built and it is being tested. It will be the main aircraft of the air force. There will also be one training aircraft if MiG and Yakovlev-Dondukov design bureaus come to terms'.[16] At a later date, there will also probably be 'scientific research studies to determine the potential capabilities of the multi-role fighter aircraft, the most advanced aircraft of the early twenty-first century to accomplish the support of the troops[17] (within the framework of existing financial possibilities).'

Nevertheless, Russia today lacks the forces necessary to provide a conventional defence against, for example, a hypothetical Chinese attack to the Far East. And as a result of the military technology transfer from Russia to China,[18] some in Moscow, such as Alexander Sharavin, director of the Institute for Political and Strategic Analysis, warn that the People's Liberation Army (PLA) is rapidly becoming more battle-worthy than the old-fashioned Russian army. Furthermore, for Russia, the use of its present tactical nuclear arsenal, even in a major military conflict with China, is very unlikely because of the proximity of almost all major Russian cities and military headquarters in the region that has a common border with China. The use of non-strategic nuclear forces could become possible only if Moscow would use longer-range tactical nuclear weapons that would threaten China's hinterland and major cities beyond the common border. Recognising these defence dilemmas on its potential eastern front, Russia may develop a new generation of low-yield tactical nuclear weapons and munitions, delivered to targets by both strategic and tactical delivery systems such as the newly developed Iskander 400 KM short-range missile.[19]

In addition to conventional weapons, Russia exports missile and nuclear technology. This has long been a source of disagreement between the USA and Russia, since the White House views those transfers with greater concern, mainly due to the current situation regarding Iran.[20] Russia has also received repeated warnings from the USA about the dangers of enhancing China's military capabilities. However, for Russia, China is for the time being viewed principally as just another source of hard currency.

In general, the Russian Military Industrial Complex (MIC) has the capability to undercut western suppliers because its prices, as in Soviet times, are unrelated to the true costs of production. There are no exact figures of how much Moscow spends in direct subsidies on the country's some 1,700 state-linked defence enterprises. Combined with indirect subsidies such as artificially low prices for energy and rent, the true cost of manufacturing might actually be higher than any profits the country receives. Furthermore, the Russians are not reluctant to export even some of their most advanced weapons, many of which are of extremely high quality. In defence of this policy, Russians argue that the proceeds from these sales offer the only way to fund the development of the next generation of weaponry.[21]

The Russian MIC actually has a vast excess capacity, but is undertaking little productive work. Even its production for the civilian sector is falling as the goods it makes are not of sufficiently high quality to compete with imports. Writing in *Segodnya* on 7 October 1998, Pavel Felgengauer argued that 'the Russian MIC can survive only as a small, separate, narrowly specialised sector. If [Russia] attempts to continue the Soviet tradition of combining the development and production of TV sets and teleguidance for aviation at the same firm, then televisions will spontaneously explode as they used to and half the bombs will miss their targets.'

Experience shows that the factories are in no condition to start new production aimed at competing in civilian markets. This is in spite of the fact that the basic stock of machine tools is adequate and, in several cases, even modern. The problems lie elsewhere, since there are people among the managers, engineers and workforce, even at the highest level of government, who have no intention of abandoning the old Soviet system with its overemphasis on research and development (R&D), design institutes and factories.

In contrast to these old voices, the Putin administration, in general, pictures a totally different situation. Its military-technical policy is focused on R&D and the creation of new models of military hardware, envisagins a situation in which scientific and designing structures will have to pay for the maintenance of idle production facilities within holdings. As a matter of fact, since Russian technological capabilities might be unable to cope with the task of producing new weapons because of a lack of capital,[22] the market of military hardware may be replaced by the market of ideas and know how[23] (at least until the necessary financing is found).

Despite all of the above elements, one should not get the impression that Russian forces have never achieved an RMA. Soviet forces were trying to keep pace with their American counterparts and, in some cases, they have even taken the lead. And after the difficult decade of the 1990s, the country started stepping up for one more time. For example, through the implementation of President Putin's military doctrine, Russia managed to develop weapons systems, such as the aforementioned mobile launched-based Topol-M2 ICBM or SS-27, a type of weapon the USA does not possess or has not yet allocated funds to develop.

Five years after his designation, on 17 November 2004, Vladimir Putin said that Russia would soon deploy new nuclear-missile systems that would surpass those of any other nuclear power. 'We are not only conducting research and successful testing of the newest nuclear missile systems', he said. 'I am certain that in the immediate years to come we will be armed with them. These are such developments and such systems that other nuclear states do not have and will not have in the immediate years to come.' The type of the system under examination was not revealed, but the Russian military is reported to have been trying to perfect a variant of the Topol Iskander-M, or Boulava, missile. Some say that the Topol-M could become operational before 2006.[24]

Russians are also working on the creation of a new generation of nuclear warheads as well as a new stealth bomber and the 5,000-km range stealth 101 cruise missile for long-range aviation. Work is also ongoing on the development of the fifth-generation Borei-class ballistic-missile submarines, a new submarine-based ICBM, the Acula-2-class nuclear attack submarine and many other weapon systems.[25] Air-to-surface Precision-Guided Munitions (PGMs) are being developed in order to add to the effectiveness of air support to the ground forces. Their current lack of an all-weather capability, and their enormous expense, might limit their use to attacks on only the most important targets.[26] However, Russian experts appear to be working towards the solution of this problem, as they know that superiority in the RMA proceeds mostly from superiority in information weapons: Reconnaissance, Surveillance and Target Acquisition (RSTA) systems, and intelligent C3I systems.

Consequently, at the time of writing, since 2001, more than 30 successful missions have reinforced the ageing Russian fleet of satellites. It is quite symbolic that the first three launches of the new space programme of the Putin administration were: a Kobalt imaging satellite, a Parus satellite belonging to the Tsyklon-B navigation and communications network – which supplies critical information for high precision warfare – and a Molniya-3K satellite for military communications. Subsequent missions have successfully delivered several types of satellites, among which are some Raduga-1 (communication), US-PU (designed to provide electronic intelligence and missile guidance for the Russian Navy), Gonets D1 (parts of a low orbital communications network), Yantar (imaging and reconnaissance), Oko-type (early warning), Tsikada (navigation), Araks and Molniya (observation and communication), Neman and Don-type (low-resolution imaging surveillance) and some Condor-E (surveillance).

These apart, the Glonass System has received a considerable amount of effort and a significant budget. Within this framework, several Uragan and Uragan-M satellites have been launched. The Glonass network has initially been designed to include 24 satellites evenly spread over three orbital planes. 'Purposeful use of money will make it possible to qualitatively change the parameters of the system so that it locates diverse objects at different points of the world with high precision.'[27] Due to lack of funds, only eight satellites were functioning by the end of 2003. As a result, the Glonass network was able to provide less accurate navigation than a completed system. However, new-generation Glonass-M satellites are currently being launched. For example, the Glonass-M 12L was launched on 25 December 2004 and another was programmed to be launched in mid-2005.

At the same time, on the ground, several types of surveillance, guidance and target tracking radars are being developed and updated. The Kasta-2E2, a world leader, is only one of the examples. Others follow it, such as the GAMMA-DE, the LEMZ 76N6, the S-300PMU/PMU1 and the PMU2 Favorit (SA-10e), the ToR-M1 and the famous NEBO series (NEBO 55G6–1 and NEBO-SV 1L13–3). Underwater, the MG and MGK series of Sonar and the Shkval torpedo complete the spectrum of Russian RSTA systems.

Putting aside RSTA, the Russian military has long believed that EW has become a form of defence against precision weapons and advanced C3I systems, by being capable of blinding the electronic equipment of opposite reconnaissance and air defence systems. It has therefore become a necessary element at all levels of the military art. To respond to this need various EW systems have been developed. The SPN-2 and SPN-4 high-powered ground jamming stations and the AKUP-1 automated jamming command-and-control complex are among the most striking examples of these systems.

Apart from new hardware production, and as a result of the rapid technological advances, Russia recognised the compelling need to control the speed of change in information technology. For the Russians, the initial concern regarding IT has been its possible impact on society, and on the strategy and tactics of the armed forces. Soon enough, IT climbed closer to the top of the list of Russian strategic priorities, since the Russian officials understood that real military power would in the future not only be determined by the quantitative, but also by the qualitative parameters of the force, 'allowing for the implementation of IT to achieve interoperability in planning; to integrate technical systems that support command and control and logistics functions; and to successfully utilise indirect actions to supplement direct deployments and strategies'.[28]

The element that was easily understood was the fact that IT can simply enhance the military effectiveness of weapons systems, being the most cost-effective way to increase combat capabilities without actually increasing weapons' quantity. It raises, for example, the combat potential of precision weapons and affects correlation of forces calculations since it can give the ability to hit strategic targets from anywhere via cruise missiles. This made further clear the fact that quantitative and qualitative indicators of weapon effectiveness have been replaced by the amount of informatisation a weapon contains. Besides, IT helps overcome uncertainty in war and limits surprises, a fact that by itself changes the art of war.

In order to face those issues, in September 1997 Russia's security council discussed the draft version of the country's information security policy. Immediately afterwards, Russian armed forces started working on combining IT with older psychological concepts such as reflexive control[29] and using IT to develop virtual realities and synthetic environments in military affairs. Soon enough, Russian IT scientists focused on the use of IT to help integrate new weapons systems to everyday military practice. Two of the main uses that were developed were the virtual reality training of Russian officers and an emphasis on testing weapon systems through virtual reality means, before they were acquired. Finally, virtual reality also began to be used by the military leadership to improve doctrine and test personnel and equipment loss according to varying climatic conditions, times of day and levels of readiness.

However, IW ('informatitsionnoye protivoborstvo' in Russian) has always had a dual importance for the Russians, since, apart from its information/technical significance, the information/psychological aspect of IW has never been abandoned. The creation of the Federal Agency for Government Communications and Information (FAPSI) indicated this duality, as the two main domains of its expertise were computer viruses[30] and PSYOP operation elements. FAPSI immediately became one of the primary operators of SOUD (the 'Russian Echelon', as some term it). In September 2000, President Putin approved the Russian Federation Information Security Doctrine and, three years later, he abolished FAPSI by a presidential decree, dividing its functions between the all-mighty FSB and the Ministry of Defence. All these elements indicate the mounting importance that IT is obtaining in the minds of Russian officials.

Nowadays, small numbers of computer-aided troop and weapons control stations have been integrated into the Russian army. End-users as low as battalion staffs or individual soldiers in the field now have the possibility to use such technology. Two areas are being further developed: the first is that of non-lethal, impact weapons for troops currently employed in peace operations and the second is that of 'functional destruction means' weaponry that can serve as a deterrent for high-precision weapons.

Finally, the timely gathering and utilization of information is being seen as increasingly important. The integration of the information gathered from reconnaissance and C2 equipment is a critical element of the Russian combat systems theory. For the military, the goal is now to develop those elements that permit to integrate the gathered information quickly into systems requiring constant data links for accurate responses. Creating everything from scratch is not the only way Russian officials are envisaging. IT acquisition is also accepted since it represents the fastest way to catch up with the West and increase combat capabilities.

10.3   Conclusion and evaluation

It was Boris Yeltsin who first advanced the idea of switching to a professional army. In autumn 2001, President Putin called for Russia's conversion to be realized by the year 2010. Many experts doubt whether this timescale is realistic, as the likelihood of improvement hinges on two challenges that must be overcome: low levels of available funding; and military resistance to reform. The military, especially the top generals, compound the problem as they are opposed to reform proposals because any such proposal, according to their perceptions, entails that the military's power be diminished. 'Our generals were trained in the Soviet times, and think the Soviet Army was the world's greatest', says Pavel Felgenhauer. 'Their aim is to restore that army, not move forward to a new type of force'.[31]

In April 2002, Marshal Sergeev, an advisor to Putin, said that the transition to a contract army needed to happen at the same time as a programme to modernize the army, especially communications, intelligence, strategic nuclear forces and the development of a space army. The fact is that to date the possibilities offered by the Russian regular and military economies have not been great. Defence spending and procurement appeared oriented towards nuclear war scenarios and supporting more R&D to exploit the RMA – for example, new mobile-based ICBMs, SLBMs, investments in strategic anti-submarine weapon (ASW), R&D in conventional and strategic C3I systems and new fighter planes.

Thus, since internal procurement has been impossible to date, the MIC has been permitted to export even state-of-the-art systems, with practically no state restrictions. Russia's putative rivals, or their own regional rivals (China, India, South Korea, Indonesia and Iran), can therefore obtain high-class weapons and systems, relatively cheaply. They can also compel Moscow and other suppliers to offer them offsets to build their own weapons and thereby further reduce seller's leverage over them. Although many of these states are Russia's potential enemies, the government sees no conventional or nuclear threat at the higher end of the spectrum of warfare for the best part of a decade.

Apparently in the same framework, on 16 October 2001 Putin announced the closure of Lurdes, Cuba's military complex, Russia's largest military base and electronic listening post in the Western Hemisphere. Established in 1964, the 28-square-mile installation housed 1,500 to 1,600 full-time personnel and was operated jointly by the 6th Directorate of Military Intelligence and the federal agency for government communication and information.[32] Annual renting costs amounted to around US$200 million (3 per cent of Direct Defence Expenditure), excluding the salaries of the base's personnel. According to Russian officials, a similar amount could purchase 20 reconnaissance satellites or 100 sophisticated radar stations. On the same day, Putin stated that Russia would also shut down its listening post and naval base in Cam Rahn Bay, Vietnam, which had been open since 1979. In any case, these two facts, along with the new international situation after September 11, are indicative of both the change in the political orientation of the Russian government, and also of a greater comprehension of the state's economic and technical priorities.

With regard to ground forces (SV), redundant structures and duplication in C2 mechanisms are being eliminated in order to minimize decision-making time and provide for a more rapid response to quickly evolving situations.

However, according to some analysts, Russian R&D, rather than being conducted in a comprehensive, scientific and practical manner, pursues several narrowly specialized directions, which have little connection to Russian combat capabilities or the missions that the Russian armed forces might be called upon to perform. To cope with this problem, the Russian officials are planning to maintain the structure of C3I elements in conformity with the modern requirements, by extending service life of technology installed in them and fitting them with advanced assets, while simultaneously creating the necessary amount of R&D,[33] for the improvement of the infrastructure. 'The modernisation programs should be aimed not as much at acquiring new weapon systems, as on upgrading the C3I systems and investing in basic infrastructure.'[34]

As we conclude, the Russian general staff continues to plan for a future war based on RMA. In the short term, they seem to be pragmatically oriented to stop-gap solutions and operational countermeasures (including 'non-traditional' weapons[35]). Whereas for the long term they seem to be trying to create an infrastructure that would permit rapid production of state of the art prototypes. For the transitional period between the two they seem to rely on their nuclear arsenal.

10.4 Perspectives

As a result of limits on defence spending and many other problems, the Russian military has been through a period of considerable upheaval since the collapse of the Soviet Union. Is financing the final driving factor for the Russian RMA?

Over the last few years the Russian military has been funded at around 2.6 per cent of the Russian GDP of $308 billion.[36] This meagre budgetary outlay was clearly insufficient to maintain a battle-worthy army, much less to fund the expensive projects necessary to reform the armed forces (especially R&D). However, recent changes in the West's policy towards Russia and rapprochement with the USA have given Moscow an unprecedented chance to cut costs and free up funds to restructure the Russian military. These new western economic and political benefits, being favourable to Moscow, seem to allow Russia to concentrate on military reform and transform its ground forces from their recent troubles into a modern army, whilst funds from arms and an increase in oil revenues will keep the defence industry alive and support R&D.

In March 2002, Defence Minister Sergei Ivanov, Putin's closest confidant in the government, said that spending levels, including procurement of 'big-ticket' items such as nuclear submarines and jet fighters, would not increase 'for a few more years. Instead the government will work to repair its decaying network of intelligence satellites, increase security at nuclear facilities, outfit existing squadrons of planes with new engines and try to modernize with precision-guided munitions and IW systems as it can afford them.'

Despite the rhetoric, Russia's direct defence spending increased significantly over the course of that year, the additional funds being almost entirely used to buy new equipment. This expenditure represented the biggest sum for a decade to be allocated directly to new armaments procurement and R&D work financing. As to the reason for the beginning of such urgent rearmament, Putin said: 'the situation in the world is forcing us to do this'.

Indeed, the focus of new procurement spending is directed towards the modernization and upgrading of air and naval weaponry and, in particular, satellites, high precision weapon systems and reconnaissance-strike complexes. However, their production will probably not begin until after 2006 or even later.[37]

Of course, money aside, what is also needed is better military-technical cooperation within the Russian Ministry of Defence. New technologies may exist, but their military applications are not yet apparent. An RMA can only occur through strategy, and strategy, starting from the very higher levels of government, should go all the way down to the last soldier. Sophisticated weapon systems should be sold only after approval of the president, the government and the Ministry of Defence. Their tasks should also include the prevention of the illegal export of defence technologies and uncontrolled supplies of individual weapons to other countries. They should take into consideration the state's interests in military-technical cooperation, as well as the fact that any agreement on such cooperation could have an impact on the country's international relations.

Care should also be taken of the staggering level of financial waste. As in Soviet times, Russia's official defence budget represents only a portion of the money that Moscow spends on the military. Much spending, including military construction and arms production, is concealed within the budgets of various ministries, state committees, and an extensive network of semi-contractors. Those weapons are then sold to other countries in what Stephen Blank, of the US Army War College, describes as pyramid-type schemes, where funds obtained from one source are used to cover short-term debts elsewhere. Consequently the military budget must become increasingly open and detailed. The entire military budget, with confidential items being the only exception, must be openly discussed in the Duma. Moreover, once approved, the budget should be strictly observed by all army and government officials, while the state Duma and Federation Council must have the right to check the ways money is spent on an itemized basis. When this is achieved, the defence orders will be fulfilled, R&D work will be paid for on time, the armed forces will get new equipment, while officers will get their pay.

In reality, in the new future Russia is unlikely to possess the economic and technological resources required to match the USA in advanced military technologies. This deficiency may force the general staff to continue relying on more territorial and brutal solutions, most notably the employment of nuclear weapons. Another interim solution would be to develop asymmetrical or niche technology capabilities[38] to create 'an electronic Pearl Harbor', or biological weapons to target the human biosphere.[39] Finally, a third short-term solution would involve the development of the Russia–China nexus, in a way that it would transcend the simple cash-for-weapons transactions of the past and form the basis of a much more unpredictable alliance in the future.

One thing is clear: despite their economic malaise, the Russians are obsessed with the aim of competing militarily in the future as they have done in the past. The difference appears to be that they are establishing priorities now, whereas in the 1960s and 1970s multiple programmes across all of the services received priority.

Rearmament of the army will have to begin this year (2005); the intention is to update 5 per cent of weapons each year and to complete rearmament until 2025, placing an emphasis on those areas where Russia retains competitiveness.[40] The means are there: the 2005 military budget, for example, is three times the level of the 2004 budget.

In order to attain their international objectives, Russian military officials have already developed both interim and long-term strategies. Will they be able to implement them? In the past they were.

Notes

1. Shurygin Vadim, 'Generals' Wars', *Novaya Gazeta*, 30 April 2002.
2. Nauchnaya-Tekhnicheskaya Revoliutsiya.
3. Col. General N.A. Lomov, *The Revolution in Military Affairs: A Soviet View*, translated and published under the auspices of the United States Air Force (Washington, DC: USGPO, 1973); John Erickson, Edward L. Crowley and Nikolay Galay (eds)., *The Military-Technical Revolution* (New York: Frederick A. Praeger Publishers, 1966); William R. Kintner and Harriet Fast Scott (trans. and ed, *The Nuclear Revolution in Soviet Military Affairs* (Norman, OK: University of Oklahoma Press, 1968).
4. Anatoly Medetsky, 'Putin OK's Plan to Cancel Conscription', *Vladivostok news*, 4 December 2001.
5. Anonymous, 'Deserters go on a Shooting Spree', *Izvestia*, 6 February 2002.
6. Like their Soviet predecessors, the Russians argue that scientific breakthroughs depend not on a country's political or economic situation, but on the brains of its scientists. Russian scientists claim to be 'still ahead of the whole planet' in many areas, and still conduct R&D in nuclear physics, high-energy and superconductivity physics, thermonuclear fusion and electronics.
7. Fitzgerald C. Mary, *The New Revolution in Russian Military Affairs* (London: RUSI, Whitehall Paper Series, 1994).
8. Aleksey Arbatov, Quote selected by Johnson's Russia List no. 5262 (AVN Military News Agency, 18 May 2001).
9. See, for more details, *Baltic Defence Review*, 6 (2001). Available at www.bdcol.ee/pages/bdr-archive.
10. Yakovlev, interviewed by Ludmila Averina, *Trud*, 13 May 2000, p. 2.
11. At the end of 2000, the Russian strategic rocket forces test-fired an SS-19 missile, reporting that it reached its target in Kamchatka from the Baikonur cosmodrome in Kazakhstan. The SS-19 missile has been part of the forces' arsenal for 25 years. The previous day a spokesman for the forces told Reuters that the missile is likely to be removed from service to join the SS-18 rocket as a booster for commercial satellites. Under the START-2 treaty, the SS-18s and SS-19s are to be decommissioned.
12. Status Report: *Nuclear Weapons, Fissile Material and Export Controls in the Former Soviet Union*, CNS Print Publication, 2001. Available at http://cns.miis.edu/pubs/print/nsr2.htm, p. 14–17, 19.
13. In October 2000 a 16-year-old Topol ballistic missile was also successfully test-fired, and a strategic Rocket Forces spokesman said that while Russia is upgrading to a newer version of the missile, the Topol-M, it will also extend the original life service of the old Topol.
14. A. Golts, 'Kremlin Moves to Rekindle Cold War Missile Plan', *The Russia Journal*, 28 June 2001.
15. The 1997 Military Planning consists of three stages: The first stage of its realization (from 1997 to 2000) called for a 30 per cent reduction in the armed forces. The second stage (2000–2005) is dealing with structural changes in the remaining armed forces, obliging no purchase of weapons and military technique and concentration on R&D. It finally foresees purchase of the next generation of weapons while the financial situation still permits. The third stage is after 2005: this should be the time for the reorganized armed forces to start being supplied with new weapons designed in the previous ten years.
16. Interfax interview with Colonel General A.M. Kornukov, 10 June 1998. The warning that no new weapons could be fielded before 2005 was given by defence

minister Igor Sergeyev in an Interfax interview on 8 February 1999: even then they would depend on defence absorbing 3.5 per cent of the GDP, he said.

17. Kornukov, in another Interfax interview, 11 March 1999.

18. The Chinese are the largest buyers of Russian arms. These purchases, amounting to well over US$6 billion in the last few years, give the Chinese the right to develop and produce some weapons themselves. This trade may well grow in the short term, but sooner or later it must provide diminishing financial returns for Russia. India is also buying a similar range of equipment.

19. See also Alexei G. Arbatov, *The Transformation of Russian Military Doctrine: Lessons Learned from Kossovo and Chechenya* (George C. Marshall: European Centre for Security Studies, July 2000), p. 18.

20. Iran is Russia's third largest arms customer (after China and India). The arms sales' agreement signed in 2001 could bring Moscow US$300 million in annual sales and could reach US$1.5 billion over the next few years; an important sum for the starved Russian military-industrial complex (MIC).

21. The Sukhoi OKB is funding the S-37 BERKUT technology demonstration programme through the sale of SU-27/Su-30 aircraft to China and India.

22. One of the main objectives of Russian military technical policy is to form a 'scientific-technical' reserve, equivalent to the western concept of 'hovering' which permits defence industries to focus on the development of prototypes and avoiding costly serial production.

23. Alexei Alexandrov, 'Restructuring and Privatisation', *Rossiiskie esti*, 14–20 March 2002, p. 10.

24. Lee Myers Steven, 'Putin Touts New Missile Advances', *The New York Times*, 18 Novermber 2004.

25. Colonel Stanislav Lunev in Maxnews.com on 4 October 2002.

26. Colonel General A.M. Kornukov, 'Win. Suppress. Support?', *Armeyskiy Sbornik*, December 1998.

27. Dubovoi Alexander, *Segodnashnaya Gazeta* (Krasnoyarsk), 20 March 2002.

28. Thomas L. Timothy, *Information Technology: US/Russian Perspectives and Potential for Military – Political Cooperation*, in *Global Security Beyond the Millennium* (Basingstoke: Macmillan – new Palgrave Macmillan, 1999).

29. A means or method used to convey specially prepared information to a person or country to influence the adoption of a predetermined decision desired by the initiator of the action.

30. Computer viruses have been considered since the beginning as force multipliers that can add new dimensions to the principle of surprise.

31. *The Christian Science Monitor*, 30 September 2002.

32. In addition to gathering and analysing US communication, Lurdes reportedly guided Russian intelligence agents to North America, provided links to the Russian spy satellite network, sent instructions to Russian ships and submarines and tracked US naval activities in the Caribbean.

33. D. Yu. Bukreyev, 'Ground Forces and Military Reform', *Military Thought*, 10(5) (2001), p. 1.

34. Baev Pavel, 'Putin's Military Reform', *Security Policy Library*, No. 6 (2001).

35. Extensive research is being conducted on laser weapons, incoherent light sources, super high-frequency and infrasonic weapons and, last but not least, Electronic and Information Warfare.

36. 'Problems of the Russian Armed Forces Must Be Addressed', *CSIS Prospectus*, 3(3) (Fall 2002).

37. See 'No Big Changes Planned in Size of Russian Arms Budget', *Jamestown Foundation Monitor*, 23 January 2002; and Alexander Golts, 'The Shadow that Lags Behind', *Yezhenedelnyy Zhurnal*, 25 (July 2002).
38. Space systems, IW, anti-satellite and RF weapons.
39. According to a Russian scientist who played a major role in the Russian biological project, there is evidence that at least until 1997 Russian scientists were continuing to work on developing genetic warfare agents.
40. See, for more details, FitzGerald C. Mary, *Russian Military Policy and International Objectives: Interim Strategies and Plans for Long Term Systemic Change*, Project on Eurasian Security (Washington, DC: Hudson Institute, 2001).

11

An Overview of the Research and Development of Information Warfare in China*

Chris Wu

11.1 Introduction

China defines Information Warfare (IW) as the neuro-system (the eyes and ears) of the services' military operation systems. China's IW encompasses C4ISR (Command and Control, Communications, Computing, Intelligence, Surveillance and Reconnaissance), electronic warfare, network warfare, and other related matters. According to the literature, the concept of IW was first proposed in China by Shen Wei Kuan, a low-ranking officer of the People's Liberation Army (PLA) in 1985. However, at that time China did not possess the necessary technological infrastructure and it also lacked any in-depth investigation into the overall IW architecture to take a lead in the research and development (R&D) of this new military scientific theory and technology. In addition, China's traditional social system and governing ideology could not adapt to the new IW ideas which had emerged within a more liberal American culture.

In 1991, during Operation Desert Storm, the US armed forces employed the concept of IW and developed a whole new set of theoretical and tactical concepts for war fighting, involving information technology and the deployment of new types of smart weapons. This alerted the PLA leadership to the fact that China needed to change its military structure by using science and technology to reinforce the army and to strengthen its air force, navy and space capabilities.

Over the course of the past decade, China has researched the theory of IW and the relevant hardware and software and has made significant advances in its technical capabilities in terms of satellite, radar, communication and Internet technology. It is difficult to obtain accurate and up-to-date information about China's IW systems due to strict secrecy. Therefore, what

---

*Note: this chapter was written in 2004 and it is recognised that Chinese technology is rapidly advancing, references to current capabilities may therefore be out of date.

follows is an analysis of the achievement of China's current and future developments based on the information that is in the public domain.

11.2   Theoretical research on Information Warfare in China

China's research on IW theory emanates from two sources – firstly from foreign countries, particularly the United States (USA) and, secondly, from a combination of ancient Chinese philosophy and war experience. However, the most important aspect of IW is the wide distribution and full utilization of information. The current inclination in China is to restrict information flow because of fears about social instability. Therefore, at present it is not easy to establish IW systems that involve the free circulation of information.

Since 1991 a number of Chinese R&D institutions have been conducting research into IW theory and technology. Amongst these are the following:

- The Military Strategy Research Centre in the PLA Institute of Science – the main IW research centre in China. The research programme at this institution includes: the development of IW strategic theory; the integration of IW into all military systems; the development of various tactics for successful IW; the participation in and organization of information technology in the international community.
- The PLA Institute of Electronic Technology focuses on the technological aspects of IW development, studying various kinds of new technologies, new devices and new components.
- PLA Joint Staff-3. The IW task is executed by the 61st Research Centre and Information Engineering School.
- National Intelligence Property and Computer System Research Centre.
- Cheng-Do University of Electronic Science and Technology.
- Shanghai Technology and Physics Department, China Institute of Science.[1]

11.2.1   China's initial research on IW theory

Although Shen Wei Kuan (see above) was the first person to introduce the concept of IW in China, he did not put forward an overall theoretical architecture. In fact, China's IW theoretical framework was proposed by high-ranking military officers after Desert Storm.

IW research in China began by studying foreign articles and analysing the television coverage of Desert Storm. Since the Chinese armed forces did not have any direct IW experience, it was impossible for Chinese military scholars alone to develop a framework for IW. They therefore copied the US concept and theory. However, there are too many differences between Chinese and US society and culture to enable everything to be copied exactly and therefore Chinese IW theory is incomplete.

As a concept of war fighting, IW is still in the development stage. As a result, the definition of IW is in a state of flux and will change in line with scientific and technological developments. While China follows the USA in its understanding of IW, it cannot lead in IW theory. Up until 2001, the state of IW in China was confused and no official national framework of guidelines or perceptive taxonomy to accurately describe the overall picture of IW in China existed.

11.2.2   A brief analysis of the US definition of IW

In the USA a broad definition of IW used by academics and the military is as follows:

1. Information can facilitate the decision making of strategic deployment and tactical manoeuvre, regardless of peace time or war time;
2. The intention of IW is to affect the opponent's decision process and benefit from it;
3. The result of IW is to cause the enemy to make the wrong decisions, delay decisions or make no decision at all;
4. The aim of successful IW is to gain information superiority over the opponent;
5. The information operation capability refers to the ability to gather, process, seamlessly distribute information, and deny the enemy of such capability;

In summary, the information operation capability defines the asymmetric strength in the field of information.

In addition, a more narrowly-defined definition of IW, in terms of tactics, is to:

1. Attack and destroy electronics electromagnetically.
2. Attack computer networks.
3. Deceive, control, distort, and provide false information to mislead the enemy.
4. Psychologically misdirect.
5. Protect and decipher encryption.[2]

In the US IW has already developed to a more advanced level, following September 11. During the Afghan war, the information system at the centre of the US command divided 650,000 square kilometres of land in Afghanistan into 1,000 small-scale IW war areas. They then used reconnaissance satellites to collect, analyse and transmit information on each area to the central Military Area Command in Florida and to carry out all-weather surveillance throughout Afghanistan. As an example of 'decision making from a thousand miles', pictures taken over Afghanistan by unmanned aerial vehicle (UAV) reconnaissance planes could be passed back to the central Military Area Command which could then command the UAV to attack identified targets with guided missiles.

After Desert Storm and the Kosovo war, the US army introduced the concept of Network-Centric Warfare (NCW). The idea was to integrate all kinds of weapons systems or platforms via a computer network in order to command and control all units from inside the network centre. NCW has greatly improved the efficiency of operations and lifted the operation of war to a higher level.

Three benefits of NCW have been considered by US military experts:

Firstly – it can enable the integration of dispersed operational forces to balance and accelerate their attacking ability and achieve a multiple force effort. In traditional war fighting, communication and the deliverability of troops and weapons is limited and so all logistic forces are concentrated on the front line around protected targets. This puts the dispersed troops at a disadvantage because it is not possible to concentrate them quickly and implement the attack. In the IW era, the expansion of sensor functionality over long distances, the increase in weapons range and the enhancement of information transmission ability can provide the separated units with the necessary supply support.

It is no longer necessary to concentrate combat troops in order to focus fighting capacity in the battlefield – thus, a revolution in military behaviour can occur. The military benefits are obvious; as the definition of fighting capacity shifts from a concentration of combat troops in front of the enemy to a concentration of firepower and information, the risks associated with fighting are reduced. In addition, the logistical load is reduced as rear-service units will not be moved frequently. The workload of health care and the transportation and supply of goods and materials – such as fuel, ammunition, and so on is significantly reduced. Finally, NCW can also enable the efficient dispersal of troops and weapons to attack a number of different targets at the same time.

Secondly – battlefield management can be enhanced. According to NCW, every soldier can see the whole battlefield and can understand the intention of the commander. In this situation, through command and control, movement of forces can be concise and coordinated and individual and joint operations can be effectively conducted.

Thirdly – every fighting operation platform is networked together. After a variety of combat units have been connected by networks in battle space, the dispersed troops and weapons can be coordinated and action can be taken more effectively. The way of fighting can also change, so that action can be taken to quickly adapt to new battlefield situations. The effectiveness will depend on the durability and high quality of the information infrastructure. There are two types of network: a 'combat system' – a hard, physical network to join combat entities together; and a soft, 'virtual procedural' network – to connect processes and procedures. The quality of the 'soft' connection is a major factor in determining how troops work together harmoniously and is the key for shaping the fighting capacity of troops and weapons. However, the effectiveness of the two types of network is determined not by how closely netted they are, but rather by how appropriate they are.

China has studied these ideas and has shifted its focus from 'Cyber Warfare' to 'Platform Centric Warfare'. Cyberwar is the destruction of the enemy's stations, information systems, computer networks and computer websites. Its targets are the computers and networks that manage this kind of warfare. The basic tactics include: attacks on computer networks, sending electronic viruses, destructive hacking, and so on. 'Platform Centric-Warfare' contradicts 'Network-Centric Warfare'. It makes the procedure of 'surveillance > assessment > decision > action' appear slow, inefficient and wasteful of battle resources. Platform Centric Warfare is focused on the platform and platform connections but not on computers and networks.

There are essential differences between 'Cyber Warfare' and 'Network-Centric Warfare' in the US army. 'Cyber Warfare' regards computers and networks as the basic tools of information. Attack and defence are defined in terms of networks and the command field in a new model of war fighting. 'Cyber Warfare' is a trial of strength in an invisible battlefield. It has the integrated characteristics of wartime and peace time and is an important method for cooperation, not only in the battlefield in wartime, but at any time.

With 'Platform-Centric Warfare' the actions are mainly centred on weapons platforms. Each platform receives and transmits battlefield information, but there is little real-time information exchange between superior and subordinate and that which does occur is slow. In addition, information sharing between platforms is very limited, which means that command officers go to the battlefields and coordinate personally by a process of 'observe, form a judgement, make a decision', which is inefficient, wastes time, does not help develop an integrated system and also reduces fighting capacity.

As described above, although NCW may have fully utilized computer network technology, it does not implement combat into the computer network. It regards computer networks as the centre and foundation of war fighting, bringing the entire battlefield, troops and each weapons platform together in an organic whole, thereby helping to improve combat efficiency. So, NCW is a brand-new model for combat – it allows information sharing in real time in all directions and at all levels, to enable 'combined' operations in the full meaning of the word, and improved combat efficiency.[3]

In this respect, it can be said that NCW is an expression of more advanced IW. It is an illustration of how US IW leads the world. With the associated theory and technology being constantly developed, the strategy of IW means that preparation, tactical application and technological means emerge in a seemingly endless stream.

11.2.3   The promotion of IW theoretical research in China

After Operation Desert Storm in 1991, China began to acknowledge the importance of high technology and information strength in an age of globalization and interdependence. China wants to become a major political and economic participant in the global society, in which information strength already plays a key role in international relations. Following the dramatic increase in its economic and national strength in recent years, China believes that the international community will become a multi-polar world. It sees IW from a strategic and military angle, as a remedy to the weakness of the old military system in China, offering low-cost, simple methods that allow weaker countries to deal with militarily superior ones such as the US and Japan. As to USA's strong military strength, IW may also offer a low-cost solution for the need to rapidly repair China's military strength after attack. Chinese strategists hope that the application of information technology to old military equipment will reduce the need for high levels of military spending.

Over the past 20 years, the ability to process information has grown rapidly, whilst the cost of doing so has plummeted exponentially. However, this is only one aspect of the situation. The costs of IW are surprisingly high if calculated in another manner. The costs of traditional military technology and infrastructure systems are rising remarkably. Although the cost of processing a piece of information has dropped by a factor of ten in the last 30 years, the cost of the command system is due to rise to such an extent that it will swallow the whole of the national defence budget.

Therefore, the speed of IW development in China will depend on the size of the budget that can be assigned and the range of engagement in IW development. Some projects and systems require high levels of financial support. For example, the essential space reconnaissance and related systems for the collection of information, especially the deployment of a multi-functional and omni-directional satellite network, requires a huge budget to guarantee constant widespread and long term coverage.

Chinese scholars and IW experts have also integrated their old, more traditional Chinese warfighting philosophies based on Sun-Tzu's 'The Art of War' and Mao Tze Dong's 'People War' and '36 Strategies' into IW theory. As China does not have the technical and financial ability to allow it to fully imitate the US mode of IW operations, it is understandable that it will look to develop IW theory in a Chinese style. The propositions of this style are as follows:

- In defensive IW – non-technological methods such as disguise, concealment and misinformation can be used to overcome the technological disadvantage. However, this approach may prove fruitless. For example, in the case of the Desert Storm operation, the Iraqi forces could not hide and escape cruise missile and bomb attacks guided by US reconnaissance satellites.

- In offensive IW – Chinese experts advocate 'using weak force to counter the strong force' and 'using offensive rather than defensive' strategies to pre-emptively attack the enemy commander and their control and/or reconnaissance system. This is because at present China can only launch cyber warfare and does not possess a NCW capability.

Chinese IW experts and scholars from the military and academic sectors have held exhaustive discussions about the nature, position, function, guidelines, principles, methods and practices of IW. In December 1994, China held a series of high-level meetings and large-scale seminars on an 'analysis of the national defence system and the technical military revolution' organized by the Commission on Science. This was followed by a seminar on the 'results of the revolution in military affairs' held by the organization Technology and Industry for National Defence in October 1995. The key writers on domestic and international IW questions in China are Dr Weiguang Shen, General Pu Feng Wang, Senior Colonel Baocun Wang and General Banggai Yuan.

In 1995 General Pu Feng Wang, an expert on IW in China, first proposed that Mao Tze Dong's 'People's War' could be introduced into IW. In his view the electronics, computer and information experts are the mainstay in the new 'people's war', being equivalent to officers in the battlefield in past wars. In people's minds, IW should be associated with war – war that can be carried out on home computers, mobilizing hundreds and thousands of people to attack enemy computer systems. China has a large number of outstanding software experts who have enormous potential in the field of IW.

Another expert on IW, Dr Weiguang Shen, has written that: 'The whole of society will replace the traditional battlefield. Different strata and public organizations will participate in the political activities of their country or other countries.' He suggests the establishment of an information protection army, comprising scientists, policemen, soldiers and other experts who understand IW, to defend the security of the national information domain against possible invaders.

In 1996, Dr Shen proposed for the first time that IW be defined as a war in which both sides try to grasp the battlefield initiative through controlling information and public opinion. As in the USA, Dr Shen's emphasis changes 'defend oneself, attack the enemy' into 'protect oneself, control the enemy'. General Wang also believes that the key to winning in IW is to control information.

In 1997, Senior Colonel Baocun Wang described the nature, degree and characteristics of IW. He believed that IW could be divided into three forms: type, level and characteristic, with the type of IW being defined by the trial of strength between attacker and defender during times of peace, tension and war. The levels of IW are national, tactical and strategic. Its characteristics include the command and control of intelligent operation, electronic warfare, psychological warfare, space control, hacker warfare, virtual warfare and economic warfare. Its particular aspects include complexity, transparency, limitation of target, short duration, small destruction, strong integration and a powerful commander. The procedures would include taking measures to cut off, deceive, conceal, speed up, improve and strengthen the chance of survival, and so on. Wang's analysis has made a significant contribution to people's understanding of IW in China.

Another author who has worked on the definition of IW in China is General Yuan from the PLA Headquarters of the General Staff. He gave his opinion on IW in 1999:

> He stated that IW is the struggle waged to seize and keep control over information, and the struggle between belligerent parties to seize the initiative in acquiring, controlling and using information. This is accomplished by capitalizing on and sabotaging the enemy's information resources, information system, and informationized weapon systems, and by utilizing and protecting one's own information resources, information systems, and informationized weapon systems.[4]

In 2000, General Pu Feng Wang presented a deeper and more thorough explanation of IW, in order to differentiate it from 'Information War'. General Wang believed that IW was a type of war *and* also a type of war methodology; whereas 'Information War' is a way of fighting and a fighting methodology. This mode of fighting is conducted in the computer network arena. Information War includes information detection and transmission systems, information on weapons systems and information processing and application systems. IW includes Information War, and integrates information and the ability to handle it together into a battlefield, with the information network as a playground. This definition by General Pu Feng Wang is quite close to the concept of 'Network-Centric Warfare' employed by the USA.[5]

Chinese scholars and experts have defined IW to include the following four aspects:

- Precision targeting and physical elimination. Attack the enemy at their general headquarters, command posts, C4I equipment, using smart, sneak attacks and horizontal flight weapons to implement surgical strikes, the carrier systems to include smart bombs and cannons and cruise missiles, and so on.

- Electronic war and the control of the electromagnetic spectrum. The struggle for control of the electromagnetic spectrum is an important aspect of winning the information battlefield war.
- Network warfare – computer networks can enable command transparency in the battlefield, offering real-time data, the use of reconnaissance satellites, GPS navigation systems, unmanned reconnaissance planes and global information transfer systems.
- Command and control of psychological war and deceit, including the dissemination of correct or false information to influence the recipient's mood and behaviour. The main tools are media propagation (TV and radio), distribution of leaflets, e-mail and other means of communication.[6]

As stated earlier, the definition of IW by China is very similar to that adopted by the USA. The main difference being that China has also introduced its old military philosophy as expressed in the 'Art of War' by Sun Tzu and the military concepts of Mao Tze Dong's 'People's War' into its IW theory.

Can China incorporate its cultural philosophical thought according to its national traditions and military situation and thereby succeed in creating its own distinctive style of IW theory? At the time of writing, this is still a difficult question to answer. Currently, IW development in China can be likened to the situation 40 or 50 years ago when it began to develop nuclear weapons. At that time China followed the same path as the Soviet Union, taking on its basic theories and technological structures and then looking at US nuclear technology, although the basic framework remained Russian. This methodology did not bring China to the forefront of nuclear technology development.

Initially, China developed its IW theories from US models and employed US technology and equipment. Hence, China finds that it is currently very difficult to overtake the USA in IW development. As the USA was the first country to develop IW theory and carried out the research and development, producing IW weapons and using them in combat during several wars, it continues to lead the world in IW. It seems likely that the current tendency in China is to develop IW theory and technology in the Chinese style whilst adopting the basic theory and technology of the US; this will mean that it will be very difficult to create a unique military science. It is most likely that China will follow other country's developments and as a result they will find it extremely difficult to catch up and overtake others. At present, there is still an enormous gap between research into IW theory and actual application in China.

It is anticipated that, in the twenty-first century, the extensive use of satellites for high-resolution reconnaissance, early warning of attack and to guide cruise missiles and many kinds of unmanned reconnaissance and attack planes, high speed, precision assault systems and the synthesis of different arms of the fighting services through a powerful information network platform, will be important components of forming IW and NCW. However, these are the very aspects of the Chinese armed forces that are least developed at present. Although efforts over the past decade have produced some IW systems and new weapons for the armed forces, they do not meet the demands on performance and quantity that war would impose. In addition, the Chinese armed forces do not currently possess NCW ability.

11.3   Current IW development in China

After settling on basic definition of IW, China has invested heavily in manpower and financial resources in order to develop it further. In this respect, considerable achievements have already been made in new weapons development and the establishment of an army network and training.

11.3.1   System and system building

In 1999 the Information Engineering College, Metrology College and Electronic Engineering College were merged with the PLA Information Engineering University. This university became responsible for the technical development of IW and engages 13 academicians as a 'brains trust'. The academic structure of the teaching team has been upgraded to include a technological officer trained in key technologies in modern IW, with particular strengths in the areas of information safety, communication engineering, space information and architecture, and so on. Courses are offered in the specialized subject areas of electronic engineering, information engineering, network engineering, control engineering, and electronic warfare. A first-class multidisciplinary research group has been set up to carry out work at the forefront of contemporary information science technical development.[7]

In May 1997 the PLA Headquarters of the General Staff (under Minister Youcai Zhang) joined with four established information leadership committees to hold a seminar on 'IW and electronic warfare'. Here it was proposed that 'information means the capacity for combat' and the concept of IW was divided into 'strategic IW' and 'IW for National Defence'. A leading group for IW was established to guide research and development for the PLA.[8] Civil resources were used to develop the military research facility. CNN's senior analyst on Chinese issues, He Li Lin, published an article saying that a military research institute had been established in Beijing to pursue military modernization and the expansion of military power in China. A diplomatic news source from Beijing reported that former President Zemin Jiang had sanctioned the establishment of a trans-departmental organization to coordinate research and development in military science and technology from 2002. This organization comprised officers and experts from the Commission of Science, Technology and Industry for National Defence, the Chinese Academy of Sciences and the Chinese Academy of Engineering. The research institute will cooperate with the General Armaments

Department of the Chinese PLA, and other similar organizations, with the intention of developing advanced technology and hardware.

There are several aspects to this development. The Chinese Academy of Sciences and the Chinese Academy of Engineering are civil organizations, rather than military units. Although most of the business for the Commission of Science, Technology and Industry for National Defence comes from the PLA, it is classified as a department of the State Council. Thus, the PLA will be relying on the resources of civil organizations to serve the military for a long time. The function and status of the Chinese Academy of Sciences and the Chinese Academy of Engineering have been improved through a series of new appointments. The former Shanghai mayor Guang Di Xu was appointed as the president of the Chinese Academy of Engineering.

It is an open secret that, in 2002, the 166 billion Yuan budget of the Chinese PLA was only equivalent to about a third of the actual level of military spending. Although the office of the Chinese Ministry of Finance stated at the National People's Congress that China does not possess a special or hidden military budget, western defence analysts believe that the money for research and development of advanced weapons systems is sometimes actually diverted from scientific research institutes or from some other civil budget of the State Council.

In October 1998, the military region of Shenyang held a seminar on the PLA definition of IW, discussing its characteristics, positioning, function, guidelines, principles, style, command, coordination, logistics, battlefield management and training. Experts from the Mechanized Infantry Division (81178 unit) have predicted that information attack and defence are likely to combine in the future. Information attack would include electronic and network attacks, element destruction, psychological attack and misinformation on military affairs. Information defence would include information retaliation, protection and resumption.[9]

China has paid particular attention to the wars in Kosovo and the Gulf and has translated relevant US documents, including the Pentagon papers on 'FM100-6 Information Combat' and the 'JP3-13 Information Combat Coalition Textbook' – standard IW teaching materials.[10]

The first 'computer and information assurance regulation' has been published by the PLA and approved and signed by the CPC Central Military Committee. This newly issued publication includes 41 regulations and is on the same basis as the 'security regulation of the Chinese PLA' and the 'technology safety security regulation of the Chinese PLA'.[11]

The Chinese IW defence and offence system was developed and successfully evaluated by the Chinese National Defence University. As a popular issue in international military studies at present, the network defence and offence system is the key technology for capturing control of the information power base and winning the high-technology war in the future. According to a report from the *Liberation Army Daily*, after four years' hard work, a research group at the Chinese National Defence University has been successful in developing the latest theoretical model for network systems offence and defence.[12]

China was the 71st international member of the Internet and, at the time of writing, has established five key computer networks at home, namely CSTNet, the Chinese public computer internet, the Chinese education and scientific research computer net, the Chinese Golden Bridge Information Net, and the CHINAUNICOM Internet.[13]

On 16 January 2003, the China Internet Network Information Centre (CNNIC) announced the 11th 'statistical report of the state of Internet development in China'. The report showed that, on 31 December 2002, the number of netizens on the Chinese Mainland had increased by 75.4 per cent on the previous year to 59,100,000 – second only to the USA. The report also estimated that this number will increase to 86,300,000 by the end of 2004. According to the report, China's netizens account for only 9 per cent of the global total and for 4.6 per cent of its total mainland population; meaning that there is still scope for huge growth.[14]

The Computer Institute of Wuhan University set up a bachelor's degree in 'safety of information' and enrolled new students formally in the autumn of 2001. It was approved by the Ministry of Education of China and at the time of writing is the only university bachelor's degree in safety of information in the whole of China. Wuhan University says its computer department has offered a course on computer safety and privacy since 1984. There is also a research group specializing in the field and relevant subject areas. A member of the Computer Institute of Wuhan University claimed that the degree course intended to recruit 50 students in autumn 2001, up to 100 people in 2002.[15]

11.3.2   The development of new weapons for Information Warfare in China

The following projects indicate that China is moving closer to the development of new IW weapons:

(a) *Synthetic Aperture Radar (SAR)* – China has pioneered multi-modal microwave remote sensing which can operate in all weather conditions – even when 60–70 per cent of the sky is covered by clouds. SAR uses the long-range propagation characteristics of radar signals and the complex information processing capability of modern digital electronics to provide high-resolution imagery unconstrained by light or atmospheric conditions. An all-weather reconnaissance and photographic system, composed of microwave radiometer, radar altimeter and radar scatter detector, was launched with the fourth Chinese spaceship or 'Divine Vessel'. After six months in orbit the remote sensing device was still working properly and the accumulated material could provide the technological foundation for ocean and meteorological surveillance satellites. The PLA's use of SAR on satellites is seen to increase tension with Taiwan and the USA.[16]

Chinese Academy of Sciences, Shanghai Institute of Technical Physics – currently, there are six Chinese satellites in space built by the Chinese Academy of Sciences, Shanghai Institute of Technical Physics. On 25 March 2002, China launched its third spaceship with five scientific devices on board, including a medium-resolution image spectrocomparator for taking high-quality pictures. The marine water colour and temperature scanner on China's first marine satellite can not only monitor the density of mud suspension, water colour, and so on, but can also determine the marine temperature.[17]

China is the third country to possess a satellite navigation system – at the end of 2001, China launched its second navigation satellite, 'The Big Dipper', from the Xichang satellite launch centre and become the third country to have a satellite navigation system following the USA and Russia. The system consisted of two satellites to cover all Chinese territory and offer accurate navigation for highway and railway transportation. It also provides a navigation service to Chinese merchant shipping in the Pacific Ocean. In addition, satellite navigation can play an important role in national defence. According to reports, at the time of writing a second generation Chinese navigation system is in the planning and design stage.[18]

Small killer satellites – China has continued to advance its space capabilities in recent years, with manned missions and the development of reconnaissance, communications and navigational satellite systems. All space-based systems are vulnerable to attack and the USA, Russia and China have all considered (and in some cases tested) anti-satellite (ASAT) systems. An article in the *Tsingtao Daily* of Hong Kong in 2001 reported that Beijing had developed a kind of 'parasitic microsatellite' that can stick to large-scale satellites and destroy them.[19]

The ultra fast electric big gun – A new Chinese concept weapon, the 'ultra fast kinetic energy electric big gun', has now reached its final stage of development and is to be tested in the near future. There are two kinds of electric big gun – the 'electromagnetic big gun' and the 'electro thermal big gun'; both are referred to as electromagnetic launchers in their non-military applications. The electromagnetic big gun is a strategic weapon that utilizes an electromagnetic force to expel a warhead to an ultra-high velocity of 50 kilometres per second, which conventional weapons are too slow to match. The electro thermal big gun uses electrical heat energy to expel a warhead with a maximum exit velocity of about 3 kilometres per second and can serve as a tactical weapon. Electric big guns can launch a few grams of various grades of small bullets or tons of warheads which greatly strengthens its penetration and its power of attack. They can also deploy more than one warhead in multi-stage launchers at the same time, thus greatly increasing their combat effectiveness. It has been reported that the electric big gun would be deployed on the ground, at sea, in the air or in space. According to news reports, research into a space-based magnetism big gun has also made good progress.[20]

Chinese 'Windows' – A technological assessment of 11 January 2003, led by Cheng Wei Wang, an academic from the Science and Technology Committee of PLA General Armament Department and He Quan Wu, the vice-president of the Chinese Academy of Engineering, indicated a crucial breakthrough in the 'dragon-core' CPU research programme.[21]

Super Computer Development – A press release from the Chinese Legend Group issued on 30 August 2002 announced the development of the 'Legend SenTeng 1800 super computer system' in Beijing. China therefore became the third country to develop a super computer, after the US and Japan.[22] The super computer is an important component of the Chinese military space network system. The latest super computer 'Dawn' 4000 A15s installed in Shanghai as a super Router ranks the tenth fastest in the world.

Operating Systems Development – On 28 September 2002, the Chinese Academy of Sciences, Technical Institute of Computation Science issued the latest edition of a high performance general CPU.[23] Seven major IT groups (Institute of Computing Technology of the Chinese Academy of Sciences, Haier Group, Great Wall Software Company of Great Wall Group, CS&S Ltd, CAS Red Flag Ltd, Dawn Group and 'Divine Boat' dragon core) joined together to make 'dragon core' – the first IT industry chain to form in China.[24]

Cruise Missile Development – On 12 January 2000, *Jane's Defence Weekly* published a report about cruise missile development in China, together with a picture of cruises missile with the serial number X-600. China tested its first turbine driven cruise missile (called 'Red Bird' and similar in type and appearance to Russia's Kh-55) in 1985.[25]

The projects listed above show that China has developed systems that could be used in IW. However, although China has spent a great deal of time and money and made significant achievements, it is still a long way behind the US.

11.3.3   Establishing network troops

China has established IW bases – The US military have identified five IW bases where they are gathering large quantities of high-tech talent, and training their own IW troops. The research claims that this reserve electronic army is now large enough to become a major force with a fighting capacity that can conduct war operations across continents.[26]

11.3.4   Chinese IW strategy

The US military claims that the IW strategy of China uses the first five of China's traditional 'thirty-six stratagems':

1. Cross the sea under camouflage – reduce the enemy's vigilance with deception. For example, hide a virus in routine e-mail or network services.
2. Besiege Wei to rescue Zhao/Surround one state to save another: Do not attack the enemy with a nuclear weapon because of the possibility of destruction by the retaliation. Instead, attack the enemy's treasure – such as the financial system, government organs, and so on.
3. Kill with a borrowed knife: Even though you may not be strong enough to attack the enemy directly, you can make use of the strength of others. For example, launch a virus or send false information via a third party.
4. Wait for the enemy: Consume the enemy's combat strength and then attack them. From the theory of 'people's war', launch a lot of attacks, but not the main force, keep the opposition teams constantly on the run, then the main force can be launched into the final decisive battle.
5. Loot a burning house: For example, a hacker pretending to be a student or a businessman can slip into and destroy the opponent's computer systems and steal information resources.[27]

11.3.5   Difficulties with developing IW in China

Although China has managed to achieve important developments in the theory, strategy, tactics, weapons and computers in respect of IW, some basic weaknesses exist that will influence the further development of IW in China decisively:

- *China lacks creativity in software and hardware for IW development* – Most computer software and hardware is imported; IT core technology is controlled by the West; 80 per cent of the global online information resources is in English and less than 0.4 per cent is in Chinese.
- *Key hardware and core technology remains in external hands* – it was reported in March 2000 by Intel and Microsoft that the Pentium and Windows 98 include hidden software that can gain access to the online user's private data. The countermeasure which China adopted was to forbid PCs which use Pentium processors from connecting to the Internet in government departments.
- *China does not have its own operating system software* – over 90 per cent of computer operating systems in the Chinese government, army,

scientific, financial and commercial departments are supplied by Microsoft.

- *IT talent is in exceedingly short supply* – much of the best Chinese computer talent has gone to Silicon Valley. At the time of writing, there are around 3,000 people in China working on developing IT, while in the USA there are around 400,000, with 70,000 more joining every year.
- *The USA has the best IT resources* – over half of the basic equipment for the Internet is in the USA. Links with the rest of the world are through seabed fibre optic cable and Comsat in the Pacific and Atlantic Oceans. 10 of the 13 top name servers in the world are from the USA, including control by the US military.[28]

11.3.6 Threats to Chinese information security

In this age of global informationization, various types of information-based systems perform key roles for both the state and business in China. Information safety mechanisms are crucial to the safe development of information technology and the prevention of attack. The Chinese Ministry of Information Industry has pointed to six major threats to the country's information safety:

1. *Information and network security and protection are inadequate* – this is a particular problem in the financial field. In China levels of high-tech crime have risen sharply and computer crime is a growing problem.
2. *Lack of ability to check and measure imported technology and equipment* – China has no methods for screening key equipment for hidden 'Trojan Horses'.
3. *The information industry relies too much on foreign equipment* – although the Chinese computer manufacturing industry has greatly improved, a lot of key parts come from manufacturers with low production capacity and little research and development.
4. *The management of information safety lacks authority* – China lacks a unified national organization with authority to enforce information safety issues.
5. *Information crime is growing fast* – information can be stolen and relayed long distances more easily and computer crime in the areas of finance and banking is increasing rapidly.
6. *Chinese society is not aware of the importance of information safety* – it seems that people feel that the degree of informationization in China is not high enough to worry about information safety – it may be something for the future, but not now.

The safety of information is not recognized at the national level in China. If these problems cannot be solved conscientiously, the information security of China could be seriously threatened.[29]

11.4   IW tactics that could be used by Beijing to attack Taiwan

Taiwan is one of the most developed areas for Internet facilities, with a high degree of informationization. There are over 6,000 World Wide Web servers in Taiwan, most of them are business-related, and there are over two million 'netizens'. The application of the Internet involves the fields of politics, economy, science and technology, military, culture and residential daily life. The network has already become an important component of Taiwanese society, thus, should there be an IW, the entires Taiwanese society will be affected enormously. Further development of the information network means that the whole of Taiwanese society will be dependent on the network in all respects. In this way, there will be greater economic losses and greater threats in light of an IW attack. The network security and survival of Taiwan would face enormous pressures.

The whole island of Taiwan is only about 36,000 square kilometres and it is close to the Chinese mainland. It has a low defence ability and short early warning and reaction times. The restrictions of the region affect the resistance ability towards industry and weaponry system attacks.

There are four command, control, communication and intelligence (C3I) systems in the Taiwanese army at present; these are the overall system, 'Heng Shan', the 'Lu Zi' system, the 'Da Cheng' system and 'Qiang Wang'. Of these four C3I systems, 'Qiang Wang' is the most advanced, with a long-range ability to reach the mainland. The 'Qiang Wang' system is composed of a ground radar station, early warning plane and automatic command-and-control systems, and utilizes advanced computer systems to integrate radar positioning, air base and anti-aircraft information.

To date, Taiwan has set up underground and sky control centres, has perfected the networked construction with a C3I system, and has adapted the advanced American anti-aircraft radar to monitor all targets in the Taiwan airspace and mainland coastal region within a range of 463 kilometres. 'Qiang Wang' can display more than 600 targets simultaneously and can deploy 150 planes to intercept, achieving full automation of the combat process from decision to solution battle plans, assigning combat order, and command and lead the anti-aircraft interception of the target. The 'Qiang Wang' system can combine with the 'Heng Shan' system, the 'Lu Zi' system, and the 'Da Cheng' to simplify the combat command procedure and shorten the reaction time.

The 'Heng Shan' overall combat information system is established in Taipei. It is the core of the C3I system with an information-gathering and combined combat command centre. This system was established in the early 1980s and came into operation in July 1990. It is made up of combat, human resources, logistics and subsystems. It links the staff officer of the defence department with armaments, the region of war and Defence Headquarters, amongst others, through a special purpose computer, data processing and display device, and allows information transmission to command and coordinate across the whole of Taiwan.

The 'Lu Zi' system came into operation in June 1996. It applies the method of combining centralization and decentralization, the node of the system relies upon Army Headquarters, legion and Defence Headquarters. The system links each and every unit in order to share combat information. Currently, the 'Lu Zi' system is a large-scale database; it will be developed to an automatic policy-making and support system in the future.

The 'Da Cheng' system was established at the beginning of the 1980s and became available for use in May 1990. It is controlled at the centre of the Taibei Navy combat Headquarters. Its main function is to search for ocean information via command, control and communication means.

With regard to IW, computer viruses can destroy and paralyse command systems, hence Taiwan has taken appropriate measures in order to reduce the chance of virus invasions and strengthen long-range early warning abilities.

The system adopts the method of dispersion in order to prevent loss of command and control ability. In order to take precautions against hackers or virus attacks, the Taiwanese military network has already adopted a working method of entity isolation. In addition, the Taiwanese military is preparing to establish an invasion scout mechanism for simulation adaptability in order to take precautions against attacks from hackers and computer viruses.[30]

As the majority of network servers and routers are centralized in Taipei and Hsinchu, 'Heng Shan' and 'Da Cheng' C3I systems have also been set up in Taipei. If Beijing should launch an IW attack on Taiwan, its first target would be Taipei and Hsinchu with the following tactical procedure:

- Large area and high-density computer hacker attacks in 'people's war' style;
- Attack Taiwan's north and south electricity transmission and distribution network by missiles, in order to cut off Taiwan's North electric energy, and then to cut off the network of Taiwan on a larger scale;
- Release neutron bombs in the air in North, Central and Southern Taiwan, to assault and paralyse the electronic and telecommunications network equipment with a high-strength electromagnetic pulse;
- Use cruise missiles to attack Taiwan's C3I systems accurately;
- Use missiles to bring down the early warning and large-scale ground-based radars of Taiwan;
- Attack or destroy the Taiwan's reconnaissance satellites;
- Use 'earth resource satellites', small satellites and the 'Divine Boat' spaceship to spy on the US aircraft carrier group. As the US army becomes involved in military activity in Taiwan, they will be attacked by Beijing via cruise missiles or middle and short range tactical missiles.

At present, Beijing has a stronger IW capability than Taiwan. However, if the USA becomes involved in IW in Taiwan, Beijing might not be victorious. Thus, Beijing has to improve its C4ISR system at a faster rate by the end of the decade, has to accelerate development of 'The Big Dipper' satellite and small reconnaissance satellites, and has to strengthen safety and secrecy measures for the network.

11.5   Combination of US and Taiwanese resistance to IW from Beijing

It is difficult for Taiwan to defend an IW launched by Beijing on its own. If the USA and Taiwan were to join forces, the following tactics might be applied:

- Computers attacked by hackers;
- Interference, destruction and attack on China's communications, navigational and reconnaissance satellites, paralysing the function of the C4ISR system. (At the end of October 2004 the USAF secretly launched a new weapon which could be used to jam an enemy's satellite communications].[31]
- Execution of a 'Trojan horse' into China's computers to destroy their computer systems;
- Destruction of computer software through Microsoft Windows operating systems;
- Investigation of the economic, trade and financial areas of China in order to discover and utilize any weaknesses in the networks and then destroy them;
- Collection of military, economic, finance and political information through all possible websites in China;
- Disruption of external interface to the Chinese network;
- Positioning of an anti-submarine sonar network on the sea floor at vital entrance channels from the East and South China seas to the Pacific Ocean.

11.6   The possibility of IW between China and the USA

11.6.1   Information Warfare behind the airplane collision incidents of China and the USA

China's espionage and counter-espionage activities with the USA are only the tip of the iceberg with regard to IW, as demonstrated by the collision of an EP-3 spy plane in 2001. However, they do outline the future development of information competition.

In the view of US IW personnel, IW is not merely 'virus combat', but should be 'the overall military information combat' and therefore should include not just electronic warfare and latent combat, but also the traditional form of a psychological war. Ideally, all effective information in IW should work cooperatively with traditional weapons systems. The logical result of this would make US reaction times faster than those of their enemies and also enable them to view the enemy's plans.

The US military believe that the final purpose of an IW is not merely to defeat the enemy, but to direct the enemy and force them to act according to the will of the USA. In addition, IW must synthesize all military strategies carefully, and must also cooperate closely with public and diplomatic relations.

However, the fact is that IW in China is in the first stages of development, although the USA already regards China as the main challenger of – and competitor for – their hegemony in the IW field. The US military believe that China has launched a brand-new 'military information superhighway' on the basis of an optical fibre network set up in military exercises in 1998, and that this network is more difficult to detect than radio and microwaves. The USA believes that, in light of the above, China is already one of the most advanced countries in the world with regard to anti-espionage capabilities. The American military believe that this is due to China increasing their IW preparation as the result of a 'war without contact' air attack on Kosovo by NATO.

The US has also identified several weaknesses of Chinese IW. The key issue is that the majority of computer systems in China use Microsoft Windows. A high-ranking official in the US military said: 'All the countries that can pose a threat of IW to the US are using US software. In China, for example, 90 per cent of the military computer systems use Microsoft Windows and Intel chips. They know that if they attack us, we have a stronger ability to retaliate. We are the strongest and most experienced in Information Warfare'.[32]

11.6.2 The 'People's War' and 'Network Warfare'

Chinese information militia hits out with a surprise attack

The PLA has begun to establish an IW network to utilize a 'people's strategy'. In the future, if China and the USA confronted one another, the 'folk information militia' would form an IW alliance with the 'official information militia'; they would also receive support from overseas foreign citizens of Chinese origin – the 'information volunteers'. This is the only strong point and advantage in IW that China possesses.

Targeting the US military and financial systems

The PLA is currently establishing 'computer soldiers' and 'network soldiers' who are proficient in foreign languages and knowledgeable about opponent's network systems. They are familiar with diversified attack techniques, and their combat effectiveness is far above that of the current protagonists in the networking competition between China and the US. During future wars the possible duties of the 'computer soldier' of the PLA may include blocking the enemy's communication networks, permeating the network system of the Pentagon to steal relevant information, and so on. They are regarded as the 'information militia' of the 'people's war' of information which will be used to destroy an enemy's financial network system, thus incurring huge economic loss.

US strategic scholars believe that an IW attack from the PLA can be divided into two major levels. The US army networks will be attacked first, including a hacker attack on the public network systems handling logistics, communication and transportation systems. The second attack may concentrate on finance centres, telephone communications, power networks, and so on and paralyse the communication systems of political VIPs. An attack on Taiwan may also include ballistic missiles and the use of IW techniques to attack decisive military installations, and so on. The present 'war of networks' between China and the US may be the prelude of a possible future IW.[33]

11.7 Conclusion

The first battlefield for a Chinese IW is likely to be in the Taiwan Straits, and this is the most probable reason why China is currently establishing IW combat forces. Before beginning a massive attack and landing in Taiwan, they could launch IW from two directions simultaneously. The first-wave attack would be on the computer networks and electromagnetic spectrum of the political mechanisms, military facilities, transportation and energy centres, in order to paralyse the government and military at all levels. This would be followed by medium-range ballistic missile and cruise missile attacks.

Meanwhile, in order to prevent delays and attack Taiwanese reinforcements from the US army stationed in Guam, Okinawa, Japanese and South Korean bases, the PLA would initiate a computer network attack on US aircraft support systems in the region. The US army would inevitable strike back with all its strength in two ways. First, it would attack the seabed fibre optic cable that China uses to connect to the rest of the world and send out its computer network attack, and, secondly, it would also make a strong attack on Chinese IW computer systems. These attacks would be aimed at the Military Area Command of Fuzhou, in Nanjing region, the computer systems of the Beijing Military Commission of the CPC Central Committee Headquarters of the General Staff, on every military airport, naval base computer network, magazines, oil depots, railway stations and airports – all at the same time in order to paralyse them.

The USA would also probably launch cruise missiles loaded with devices to generate a powerful electromagnetic pulse to attack and destroy Chinese IW computer systems and then attack reconnaissance and GPS satellites, with the aim of completely demolishing demolish the IW eyes and ears of China. Thus, IW would be the start of an all-out war between China and the USA. Even though an army has not yet been deployed, an extensive war will have already begun. At this time the technological ability of China to conduct IW is relatively weak, thus the likelihood is that China would lose such a war.

This chapter has presented a general picture of the preparation for IW undertaken by the PLA in recent years. It has illustrated a severe disparity in the balance of forces between mainland China and Taiwan. The question is – how can this disparity be reduced in order to safeguard the security of Taiwan? From the author's point of view, two suggestions could be considered. Firstly, the Taiwanese C4ISR system and its ability to survive should be strengthened, as soon as possible, and army training and quality should exceed that of the mainland by a large margin. Secondly, Taiwan should combine C4ISR operations with the USA, and especially with their ASAT systems, in order to provide the ability to disrupt, paralyse and destroy the satellite communications, reconnaissance and navigation systems of the mainland. In this way, the current situation in the Taiwan Straits may be maintained.

Notes

1. James Mulvenon, *The PLA and Information Warfare* (Santa Monica, CA: Rand), p. 179.
2. Toshi Yoshihara, 'Chinese Information Warfare: A Phantom Menace or Emerging Threat?' Rand, November 2001, pp. 3–5.
3. Baocun Wang, MeiYu Wang, Yansheng Shi, '"Network Centric Warfare" of American Army in My Eyes', *Liberation Army Daily*, 11 October 2001.
4. Y. Banggen, *On IW, Digital Battlefields* (Beijing Zhongguo Junshi Kexue, 20 February 1999), pp. 46–51 as translated and downloaded from the FBIS website on 17 July 1999 – quoted by T. Thomas, *Like Adding Wings to the Tiger: Chinese Information War Theory and Practice* (Fort Leavenworth: Foreign Military Studies Office). Available from: http://www.iwar.org.uk/iwar/resources/china/iw/chinaiw.htm.
5. Zhanliang Wang, 'Characteristics of Chinese Information Warfare: The Point of View from Foreign Military Expert', 31 December 2001.
6. Yoshihara, 'Chinese Information Warfare', pp. 16–17.
7. *People's Daily*, 8 April 2002.
8. *Liberation Army Daily*, 16 February 2000.
9. Kove Ping, 'The New People's War of China–America Networks', *Asia Weekly*, 9 May 2001.
10. Ibid.
11. *Voice of Overseas Chinese*, 11 February 2001.
12. *Voice of Overseas Chinese*, 26 August 2002.
13. Xuewen Zhou and Xiangcheng Luo, 'The Legislation Analysis of China's Information Security in Internet Environment', *Voice of Overseas Chinese* (San Francisco: Overseas Chinese Cultural Society), 26 August 2002.

14. At the time of writing, there are 5.91 million netizens in China, occupy the second place in the world: *Daily of the World*, 17 January 2003.

15. The only bachelor's degree of safety of information of Chinese university was enrolled new students for the first time in autumn 2001; www.people.com.cn, 9 May 2001.

16. The detailed revealed of secrets for three major scientific experiments about the fourth of the Chinese 'Divine Boat' in the first time, Chinese News Network, 16 January, 2003.

17. The remote sensing technology in Chinese Academy of Sciences, 'Shanghai Institute of Technical Physics Aims at International Advanced Technology', www.people.com.cn, 23 October 2002.

18. China becomes the third country for possession satellite navigation system (GPS) in the world, *National Defence Newspaper*, 15 January 2001.

19. Gregory Kulacki and David Wright, 'A Military Intelligence Failure? The Case of the Parasite Satellite', Union of Concerned Scientists, 16 August 2004. Available at http://www.ucsusa.org/global_security/china/page.cfm?pageID=1479.

20. 'The Ultrafast Electric Big Gun of China Will Come Out', *Knowledge of the Naval Vessel*, 25 March 2002.

21. 'Chinese Military Will Develop Advanced Domestic "Windows"', *Science and Technology Daily*, 21 January 2003.

22. 'Legend Group of China Introduces the Super Computer', Central News Agency, 30 August 2002.

23. 'China First Piece High Performance General CPU-"Dragon Core" Came Out', *Beijing Evening*, 14 October 2002.

24. Zhang Xuguang, 'Seven Major IT Industry Group League to Make "the Dragon Core" – The First IT Industry Chain of China Takes Shape', *Beijing Morning*, 26 December 2002.

25. 'China Cruises Missile Plan is Exposed', *Jean's Defence Weekly*, 12 January 2000.

26. 'China Has Set Up Five Major Information Warfares Base', China.com, 26 May 2001.

27. Ibid.

28. 'China is Losing "The Power of Network Control"', *eNet Silicon Valley power*, 3 October 2000.

29. 'China Faces Six Major Respects Situation for Information Safety', *CEInet*, 9 November 2000.

30. 'Painstakingly Build Up with Limited Ability – Can Taiwan Afford to Make Information Warfare?' Hua Xia Transit Network, 17 May 2004. Available at http://jczs.sina.com.cn/2004–05–17/1643198436.html.

31. 'Great Era: New American Space Weapon System Has Started in Order to Paralyse Enemy's Communication Satellite', 23 November 2004. Available at http://www.chinaaffairs.org/gb/detail.asp?id=49285

32. Luming You, 'The Information Warfare Behind the Collision Incidents of China and America Airplane', China News Service, 23 August 2001.

33. Kove Ping, 'The "People's War" of Network Between China and America', *Asia Weekly*, 9 May 2001.

Part IV

What is Being Done – or Must Be Done?

12
A Bridge Too Far?

Mike Moore

Around noon on 7 April 2003 during Gulf War II American forces attempted to kill the Iraqi dictator, Saddam Hussein, with high-altitude bombing. The bombs were dropped from a supersonic B-1 bomber flying at close to 30,000 feet above a cloud-covered Baghdad. Special Operations forces on the ground were said to have identified Hussein and his top advisers, who were about to have a luncheon meeting in a large structure tucked into an upscale residential neighbourhood of Baghdad.

The Special Ops team immediately passed on the precise geographical coordinates of the structure to an American command centre via a satellite communications link. The coordinates were determined with the aid of a constellation of Global Positioning System (GPS) satellites orbiting roughly 12,500 miles overhead. Timing signals sent out by GPS satellites are accurate to a millionth of a second, which permits great precision in determining position through a space-age version of old-style triangulation.

Within minutes of the ground sighting in Baghdad, a B-1 bomber crew, which had just completed taking on fuel from an aerial tanker over western Iraq, was assigned a 'priority leadership target'. Some 12 minutes later, the crew put four 2,000-pound 'bunker-buster' bombs on two 'aimpoints' 50–100 feet apart. One weapon hit each aimpoint; three seconds later, two more weapons followed. Organized Iraqi military resistance collapsed soon after the bombing.

The building was reduced to rubble and an unknown number of people inside were killed, as well as several near by civilians. Was Saddam Hussein among the dead? No one really knew. So complete was the destruction that no one knew at the time whether Saddam Hussein had been killed, or, for that matter, whether he had even been in the building on that day. The world got an answer to the first question on 13 December, when Saddam was found cowering in a 'spider hole' near Tikrit.

The bunker-busters used in the attack were part of a family of bombs called 'Joint Direct Attack Munitions' (JDAMs). JDAMs almost invariably detonate within a few metres of their aimpoints, even when launched from as high as 40,000 feet. Indeed, higher altitudes tend to improve accuracy; the guidance computer attached to the falling bomb receives continuous data from the GPS system, enabling it to fine tune the trajectory of the bomb to the moment of impact. Just how accurate GPS-guided bombs are is classified, but it is measured in metres, very few metres. During the earlier war against the Taliban and al-Qaeda forces in Afghanistan, American ground forces at times called in bomb strikes to targets as close as 25 metres away from their own positions.

The new American way of precision warfare is greatly – but not wholly – dependent on linking ground-based information systems with a host of special purpose 'assets' in orbit: intelligence, surveillance, and reconnaissance 'birds', communications satellites (commercial and military), meteorological satellites and, of course, Global Positioning and timing satellites. In the military it is called 'net-centric warfare' and at time of writing only the USA has the capability to fully deploy such a system.

The synergy between ground-based information systems and space assets allows the US to apply (in Pentagon-speak) 'lethal force' rapidly and decisively. It also permits US military forces to fight wars humanely – if that word can ever be applied to warfare. If the targeting information is good, which is not always the case, the US military, whether on the ground, at sea, or in the air, is able to strike military and militarily important targets with great accuracy, thus assuring that a minimum of non-combatant civilians are killed or wounded. That is, US forces can spare civilians if the other side does not closely commingle military targets with civilians, as did the regime of Saddam Hussein.

In the Second World War, the US and the British also relied heavily on a doctrine of strategic, civilian-sparing, 'precision' bombing to help defeat Nazi Germany and Japan. Unhappily, precision as understood today was not possible with the technology then available. Millions of (mostly ignored) words have been written over the past half century about how the Anglo/American doctrine of precision bombing was transformed during the course of the war into a terrifying experience for German and Japanese civilians because true precision was not possible. The number of German and Japanese civilians who died as the result of Anglo-American 'precision' bombing campaigns cannot be calculated with certainty. The lowest estimates are in the hundreds of thousands; other estimates, some perhaps tainted by ideology or political axe-grinding, give figures of more than a million. These estimates exclude the deaths caused by the atomic bombing of Hiroshima and Nagasaki. The official US Strategic Bombing Survey, for instance, estimated that, at 'minimum', 305,000 German civilians were killed and 780,000 wounded. (Ironically, the doctrine of precision bombing was refined in the USA and Britain during the 1930s and 1940s with Germany expressly in mind.)

After the war, Air Marshal Sir Arthur Harris, who had headed the Royal Air Force Bomber Command, defended the strategy of levelling cities with massed bomber attacks. 'I never forget, as too many do, that in all normal warfare of the past, and of the not distant past, it was the common practice to besiege cities and, if they refused to surrender when called upon with due formality to do so, every living thing in them was in the end put to the sword...' (Harris, 1990, p. 177).

The American who directed the principal air offensive against mainland Japan was General Curtis LeMay, who later became chief of staff of the US Air Force. He echoed Sir Arthur regarding the systematic destruction of Japanese cities: 'There's nothing new about this massacre of civilian populations. In ancient times, when an army laid siege to a city, everybody was in the fight. And when that city had fallen, and was sacked, just as often as not, every single soul was murdered' (LeMay, 1965, p. 384).

If anyone believes that the twentieth-century American way of precision warfare, based in large measure on the use of US space satellites, is inexcusably brutal, let him or her reflect for a moment on the words of Sir Arthur and Curtis LeMay. Yet, nevertheless, there *is* a dark side to the new American way of war. America's ability to wage precision war may, in the end, help push the USA down a dangerous neo-imperial path, complete with a deployed space-control capability and, possibly, with weapons in space. That would be a first for humankind. No nation has ever sought to develop a space-control capability and no nation has ever deployed weapons in space. Indeed, an international treaty, negotiated in the 1960s, declares that space should be reserved for 'peaceful purposes'.

However, this chapter must first make something clear. Despite the fact that the new American way of war is largely dependent upon space-based hardware, the US currently has no weapons in space – no 'shooters', as they are called by those in the military space field. Indeed, no nation has shooters in space. Near-earth space has been militarized for 40 years, mainly by the USA and the Soviet Union, but it has never been weaponized.

Although a large fraction of US satellites (military, intelligence, and even commercial) are essential to the proper functioning of US weapons systems on the ground, at sea, and in the air, these satellites are not themselves weapons. That may change, however. Many influential people in the Defense Department and in hard-line 'think tanks' would like to see the US develop the capability to control space, when required, and to place weapons in space.

12.1 Global engagement

The use of space capabilities by the US military has changed significantly since the first military satellites were orbited. Continuous improvements in space technology have led to the development of more advanced space systems. Space capabilities have proven to be a significant force multiplier when integrated into joint operations. Military, civil, and commercial sectors of the United States are increasingly dependent on space capabilities, and this dependence can be viewed by adversaries as a potential vulnerability.

The United States must be able to protect its space assets (and when practical and appropriate, those of its allies) and deny the use of space assets by adversaries. (JCS, 2002, p. vii)

The key word in that ponderous prose from the Joint Chiefs of Staff is 'deny'. For at least a generation, highly placed people in the White House, in the Defense Department, and in many hard-line think tanks have believed that adversary nations will attempt to offset America's high-tech military power by attempting to disable or destroy US space assets. They assert that future conflicts will not be confined to land, sea, and air. Space will inevitably become the 'fourth medium of warfare'. Battles will be fought *in* space and *from* space. The USA had better wake up to that fact and take charge. It must pre-emptively develop the means to deny the use of space to 'bad guys'. It must have the capability to *control* space.

In 1997, the US Space Command, then an umbrella organization with its headquarters in Colorado Springs, issued a 16-page *Vision for 2020*. Printed on glossy, heavyweight paper and heavily stocked with full-colour illustrations, *Vision* looks rather like a prospectus for a gated retirement community in Florida, the sort of thing typically peppered with punchy paragraphs and salted with bromides that describe the development's unparalleled amenities (SpaceCom).

Instead of depicting a proposed championship golf course, tennis courts, pools, and clubhouses, *Vision* offers dreams of unlimited spacepower. On the first page, in oversize type against the black background of space, we read: 'US Space Command – dominating the space dimension of military operations to protect US interests and investment. Integrating Space Forces into warfighting capabilities across the full spectrum of conflict.' The type, a brilliant yellow, seems to fall away from the reader, much like the beginning of George Lucas's first *Star Wars* movie ('A long time ago, in a galaxy far, far away...').

The illustration on the inside back cover of *Vision* punctuates the *Star Wars* theme. Our vantage point is near-earth space, a few hundred miles up. Below is a pie-wedge portion of the earth, depicted in the sere sepia shades of a desert landscape. We see the easternmost tip of the Mediterranean Sea and, below it, the Red Sea, partly obscured by clouds. Above the Mediterranean is a bit of the Black Sea, to its right the Caspian Sea, below that the Persian Gulf. The rest of the painting is bluish-black space, speckled with stars. An orbiting laser dominates the foreground; it glows orange as it zaps a target on the Iraq–Iran border. Does the laser-induced explosion represent the destruction of an ascending missile? Probably. A terrestrial bunker? Possibly. The artistic evidence is ambiguous. But the didactic point is not; Space Command means to dominate space if it ever gets a thumbs-up from the White House.

Vision was a preview. The following year Space Command issued a 90-page *Long Range Plan*, a document of some importance whose publication was little noticed within the USA. One supposes, though, that the plan was eagerly read in most world capitals. Governmental officials and military officers around the world have a keen interest in trying to figure out what the USA will do next with its high-tech military power.

Like military forces everywhere, Space Command had long been accustomed to the rigours of worst-case analyses. The glass was always half empty – and it was also probably cracked. In the reasonably near future, said Space Command, all sorts of adversaries – national military forces, paramilitary units, terrorists – would acquire sophisticated space capabilities. Enemies 'may very well know, in near real time, the disposition of all [US] forces. They will command and control their forces with real-time access to precise navigation (position and timing), submeter imagery, highly accurate weather data, timely missile warning, and robust communications'. Hostile forces will 'share the high ground of space with the US and its allies'.

The plan stated that technologies available in the global marketplace would help 'bad' actors develop anti-satellite weapons. Wealthy states would probably opt for directed-energy weapons, such as lasers, to disable US space assets. 'Lesser powers' might prefer to jam signals or disable command-and-control systems and intelligence operations with cyber attacks on US computer systems.

The plan's authors peered into their crystal ball and everywhere saw darkness. They claimed that losing the use of space in a future conflict would be 'intolerable'. Developing and deploying the capability to control space during a time of high tension and even conflict would require systematic effort and heavy investment; it would be neither easy nor cheap, but it would be necessary. By 2020, the US would have a 'robust and wholly integrated suite of capabilities in space and on the ground' and it would have achieved 'dominance of space', thus ensuring the protection of US military and commercial interests.

The plan stated that the USA was in a moment of 'strategic pause'. The Cold War was history and no 'peer competitor' would appear on the horizon for at least 20 years. It was a good time to explore 'innovative warfighting concepts and capabilities'. Like airpower before it, spacepower would progress from its current role of supporting warfighters 'toward space combat operations'. Eventually, 'as it continues to mature, it may allow us to project force from space to earth' – in other words, to attack earthly targets from space.

Space control was defined in the *Long Range Plan* as 'the ability to assure access to space, freedom of operations within the space medium, and an ability to deny others the use of space, if required'. Beyond space control, the plan said, lies 'global engagement', 'holding a finite number of targets at risk anywhere, anytime with nearly instantaneous attack from space-based assets'.

US Space Command no longer exists as an independent entity. In late 2002 it merged with US Strategic Command, the direct descendant of the old Strategic Air Command. The emblem of the new and improved Strategic Command, StratCom to all, depicts a medieval knight's mailed fist, which clutches thunderbolts and an olive branch. Beneath the fist is a highly stylized Earth, encircled with orbital tracks. According to StratCom (2003):

> The lightning bolts symbolize lethality and speed while the olive branch is a constant reminder of the command's mission of securing the objectives of peace. The globe, as viewed from space, symbolizes the earth as being the origin and control point for all space vehicles and represents the command's span of operations.

The emblem's symbolic meaning is unmistakable. Call it up on the web: www.stratcom.mil. Does it inspire you with its implied vision of a beneficent *Pax Americana*? Or does it make you just a little wary?

12.2   A space Pearl Harbor?

On 11 January 2001, a report from a blue-ribbon 'Space Commission' (the Commission to Assess US National Security Space Management and Organization) was issued. The commission, established by Congress, was sharply critical of the failure of the political leadership of the USA to take bold action to protect US military, intelligence, and commercial interests in space.

The commission said that the USA must achieve the military capability 'to use space as an integral part of its ability to manage crises, deter conflicts and, if deterrence fails, to prevail in conflict'. The report contained the following stunning words, said to have been personally drafted by the chairman of the commission; a long-time Washington insider named Donald Rumsfeld:

> History is replete with instances in which warning signs were ignored and change resisted until an external, 'improbable' event forced resistant bureaucracies to take action. The question is whether the U.S. will be wise enough to act responsibly and soon enough to reduce U.S. space vulnerability. Or whether, as in the past, a disabling attack against the country and its people – a 'Space Pearl Harbor' – will be the only event able to galvanize the nation and cause the U.S. government to act. We are on notice, but we have not noticed. (Rumsfeld, 2001)

Why did a report, with such headline-friendly rhetoric, not attract more attention? Bad timing. On 28 December 2000, after the commission's report was completed, president-elect George W. Bush nominated Rumsfeld as secretary of defense. In a peculiar coincidence, Rumsfeld's confirmation hearing before the Senate Armed Services Committee took place on 11 January 2001, the same day the Space Commission's report had been scheduled for release. Rumsfeld's testimony during the hearing focused on how he would reshape the military to meet the challenges of the new century. That was *the* Rumsfeld story of the day. The findings and recommendations of Rumsfeld's Space Commission were ignored or buried.

The Space Commission's report deserved more vigorous treatment than Rumsfeld's *pro forma* comments at the confirmation hearing. It was a truly revolutionary document. Among other things, it spoke of placing weapons in space. In a nod to those who might not be enamoured of such ideas, Rumsfeld's Space Commission simply said that the commissioners 'appreciate the sensitivity that surrounds the notion of weapons in space for offensive or defensive purposes'. Nevertheless, to ignore the issue would be a 'disservice to the nation'. The president ought to 'have the option to deploy weapons in space to deter threats to and, if necessary, defend against attacks on the US interests'.

High on the commission's priority list was the development and testing of a variety of anti-satellite (ASAT) weapons, largely because the commissioners believed that unfriendly nations could deploy observation and command-and-control satellites that would someday imperil US forces on land, sea, and air.

'The senior political and military leadership needs to test these [ASAT] capabilities in exercises on a regular basis, both to keep the armed forces proficient in their use and to bolster their deterrent value.' By 'test', the commissioners meant computer simulations, war games, and 'live-fire events'. The latter would require 'testing ranges in space'.

'Earth surveillance from space' was another big-ticket item on the commission's agenda: 'The U.S. needs to develop technologies for sensors, communication, power generation, and space platforms that will enable it to observe the earth and objects in motion on a near real-time basis, 24-hours a day. If deployed, these could revolutionise military operations'.

Space-based radar aimed towards the earth, the commissioners explained, 'could provide military commanders, on a near-continuous and global basis, with timely, precise information on the location of adversary forces and their movement over time'. That ability, 'coupled to precision strike weapons delivered rapidly over long distances', would give the USA a potent new weapon to deter 'hostile action'.

In respect of the endlessly controversial matter of national missile defence, the commissioners turned cagey, presumably because it was still national policy when the committee issued its report to preserve the 1972 Anti-Ballistic Missile Treaty. The commissioners simply said: 'Some believe the ballistic missile defense mission is best performed when both sensors and interceptors are deployed in space. Effective sensors make countermeasures more difficult, and interceptors make it possible to destroy a missile shortly after launch, before either warhead or countermeasures are released.' Then came the Space Commission's money shot:

> Finally, space offers advantages for basing systems intended to affect air, land, and sea operations. Many think of space only as a place for passive collection of images or signals or a switchboard that can quickly pass information back and forth over long distances. It is also possible to project power through and from space in response to events anywhere in the world. Unlike weapons from aircraft, land forces, or ships, space missions initiated from earth or space could be carried out with little transit, information, or weather delay. Having this capability would give the U.S. a much stronger deterrent and, in a conflict, an extraordinary military advantage. (Rumsfeld, 2001. See particularly Section II: 'Space Today and the Future')

12.3   Space cop

Taking the high ground of space and placing weapons in space is hardly a new idea. Consider the 22 March 1952 issue of *Collier's* magazine, then a widely read and influential publication in the USA. The cover featured a painting of a winged spaceplane, rocket engines ablaze, bursting into the darkness of space many miles above Earth's day/night boundary. The lead article was an essay by Wernher von Braun, the Nazi scientist whose V-2 missiles had terrorized London and Antwerp during the war (Ryan, 1952, pp. 12–70).

By 1952, von Braun had become America's leading rocketman. He was suggesting in *Collier's* that the USA build a slowly rotating 'wheel-shaped satellite' 250 feet in diameter that would circle the Earth every two hours at an altitude of 1,075 miles. 'From this platform, a trip to the Moon itself will be just a step, as scientists reckon distance in space', von Braun wrote.

However, the satellite would do more than advance the science of rocketry and space travel; it would also ensure world peace. US technicians 'using specially designed, powerful telescopes attached to large optical screens, radarscopes, and cameras' would 'keep under constant inspection every ocean, continent, country, and city... It will be practically impossible for any nation to hide warlike preparations for any length of time.'

With its space platform, the USA would become space cops, armed with atom bombs rather than riot clubs. If a nation threatened world peace, 'small winged rocket missiles with atomic warheads could be launched from the station in such a manner that they would strike their targets at supersonic speeds. By simultaneous radar tracking of both missile and target, these atomic-headed rockets could be accurately guided to any spot on the earth.'

What country would go to war with the USA, or with a friend or ally of the USA, thus risking retaliation from Uncle Sam's 24-hour-a-day space patrol?

Von Braun was not along in his vision. In some military circles, space had long been thought of as the ultimate high ground. A few months after the end of the Second World War II and long before von Braun's *Collier's* article, Henry 'Hap' Arnold, the five-star chief of the USA Army Air Forces, said that in the reasonably near future the USA should develop the capability to fire atomic-tipped missiles from 'true space ships' in the earth's orbit (quoted in Emme, 1959, p. 310).

Although the USA was soon to encircle the Soviet Union with conventional long-range bombers, the space force idea never quite died. In February 1957, General Bernard Schriever, director of the Air Force missile programme, publicly asserted that the US should achieve military 'space superiority'. In a few decades, he said, 'the important battles may not be sea battles or air battles, but space battles' (Futrell, 1989, p. 549).

In the wake of the launch of *Sputnik*, Thomas D. White, chief of staff of the Air Force, launched an ardent campaign to persuade President Eisenhower to order the Air Force into space. 'The US', said White in one speech, 'must win and maintain the capability to control space in order to assure the progress and preeminence of the free nations...Control of space', said White, 'should be the goal of all Americans' (AFM, 1959, pp. 11–17, quoted in Lord, 2003).

President Eisenhower did not agree. The world was a troubled and dangerous place, he said; the last thing it needed was an arms race in space. In public, he said space should be reserved for peaceful purposes; in private, he said spy satellites were fine, even necessary. They would help preserve peace by opening a window into a tightly closed and highly militarized society. Uncertainty about the other side makes for itchy trigger fingers and preventive wars; in contrast, hard data enhances certainty.

Eisenhower established a space-for-peaceful-purposes policy for the USA that is still (barely) observed today. He also set in motion a chain of events that led to the signing of the Outer Space Treaty of 1967, which, among many things, bans nuclear weapons and 'other weapons of mass destruction' in space. Unfortunately, the treaty did not ban precision weapons in orbit, which is what today's space warriors speak of. Long-range precision weaponry was not even a concept in the 1950s and 1960s (McDougall, 1985, chapter 8).

The world of President Eisenhower and General White was long ago and far away. Four decades ago, the Soviet Union and the USA were nuclear powers and the Soviets had demonstrated an extraordinary proficiency in space. A nuclear arms race in space would have been madness. But today, the Cold War is a fast-fading memory and the USA is the world's sole remaining superpower, and, in the context of global history, a relatively benign superpower.

Given these circumstances, why would it be wrong for the USA to take control of space, 'in order to assure the progress and pre-eminence of the free nations'?

12.4 The security dilemma

More than a decade after the end of the Cold War, the world is an unpredictably dangerous place. Americans need no reminders of that. The USA requires well-trained and well-equipped military forces to help ensure its security. In any event, the USA will remain the world's dominant military power for the foreseeable future. In the words of some, it is the 'indispensable nation' (Clinton, 1997) in a chaotic post-Cold War world.

But just *how* militarily dominant should the USA be? At what point, if ever, does the drive towards achieving overwhelming military superiority inspire so much fear and loathing among other nations as to provoke countervailing reactions? Realists talk endlessly of the 'security dilemma', a zero-sum game in which a state that becomes extraordinarily powerful is seen by other states as diminishing their own relative security, thus compromising their own freedom of action.

Although there are many permutations of the security dilemma, the concept seems well rooted in real-world experience. More than two millennia ago, Sparta went to war against Athens at least in part because the Athenian empire had become so powerful as to threaten the balance of power in the Hellenistic world. The Cold War itself was an example of the security dilemma writ large; an action–reaction dynamic in which East and West sought to ensure that the 'other side' would not gain a dangerous military advantage. There was no final showdown, goes the argument made by many realists, because the nuclear powers were in rough balance. (One could also bring luck into the equation, but that is another book.)

In the post-Cold War world, America's unbalanced high-tech military superiority has inspired a growing global unease, ranging from French accusations that the USA has become a loose-cannon 'hyperpower' to Chinese assertions that the USA is intent on achieving global 'hegemony'. Much of that rhetoric is suspect. Nothing about French or Chinese history suggests that they have cornered the market on wisdom, ethics, or morality. However, the extent of America's military power *is* worrisome. Timothy Garton Ash, an Oxford scholar, a long-time friend of America, and a senior fellow at the conservative Hoover Institution at Stanford University in California, expressed it in the following terms:

> It would be dangerous even for an archangel to wield so much power. The writers of the American Constitution wisely determined that no

single locus of power, however benign, should predominate; for even the best could be led into temptation. Every power should therefore be checked by at least one other. That also applies to world politics. (*New York Times*, 9 April 2002, opinion page)

An attempt by the USA to take unilateral control of space ought to trouble everyone, including – and perhaps particularly – the citizens of the USA themselves. An attempt to achieve US dominance in space could easily evolve into a case of imperial overstretch that might, in the end, trigger events and actions that would further compromise American security.

The desire to enjoy freedom of action in world affairs is not a uniquely American aspiration. It is a universal goal of all governments, albeit are that is seldom achieved. Whether they are democratic, authoritarian, totalitarian, monarchical or theocratic, governments seek to maximize their own freedom of action vis-à-vis other states. Like naked adolescent boys sizing up one another in the locker room, regional and global powers are forever assessing their competition. The USA is, by far, the biggest fellow in the locker room.

What *does* the USA mean by 'full spectrum dominance' of the 'battlespace', words regularly used in Defense Department 'vision statements'? For citizens of other countries, large and small, the phrase may sound ominous, particularly in view of the fact that the US Department of Defense regards any spot on the globe as a potential 'battlespace' (JCS).

In assessing the threat posed by existing or potential rivals, national leaders are more interested in capabilities – demonstrated or presumed – than in intentions. Capabilities are thought to be roughly measurable. By contrast, divining the intentions of another nation's leadership is a speculative art that, in any event, is usually futile. Intentions can change as rapidly as governments.

Those of us who believe in the value of international cooperation and amity regularly condemn excessive 'realism'. It poisons the international atmosphere; one need not look beyond the Cold War for an example of that. If only the leaders of the USA and the Soviet Union had not been so short-sighted. If only they had understood the commonalty of humankind. If only they had appreciated that Planet Earth was the only home we humans would ever have. If only nations were run by saints and angels...there would have been no nuclear arms race, no continuing threat of Armageddon.

Sadly, there are few saints and angels among us, or at least those that do exist do not end up as presidents or prime ministers or rulers-for-life. When it comes to national security, the leaders of the world's nations are a suspicious lot. They drink to one another's health in bilateral, regional, and international meetings, but they keep their backs to the wall and a wary eye on everyone. Smiles and pleasantries are standard fare on the global champagne circuit, but actual trust is in rather short supply. The history of the world is written in blood and tears and the leaders of nations seldom forget that.

Assessing the military capabilities of other states is a fundamental fact of world politics. It is necessary and prudent, even though the process is always subjective, one in which facts and factoids mingle freely and are filtered through multi-hued lenses. But two facts about today's world are beyond dispute. The USA is the most militarily capable nation in the history of the world, and it has demonstrated, as have all past great powers, a fondness for unilateral action. Furthermore, its foreign policy is as arrogant at times as it is magnanimous at others. Given these realities, the USA is seen in many parts of the world as unpredictable, hubristic, and a potential threat.

'Why *do* they hate us?' is a common question in the USA. 'Are we not the good guys?' The answer is not hard to fathom. The staggering military capabilities of the USA have been clearly demonstrated in Gulf War I, in the skies over Kosovo and Serbia, in Afghanistan, and in Iraq. The USA is widely perceived as a state with the technological wherewithal – if not necessarily the intention – to do exactly as it pleases.

The USA intends to keep a couple of thousand nuclear weapons deployed for quick use and thousands more in reserve. It has a growing array of 'conventional' weapons capable of attacking targets with unprecedented stealth and precision from high altitudes. It has the world's best-trained and best-equipped military personnel and the 'lift capacity' to shuttle sizeable battle-ready contingents to any point on the globe within days or weeks. Its high-tech lead over other nations in all things military widens every year and now it speaks, with some urgency, of the need to take the high ground of space, to *control* space, and possibly to place weapons in space.

That raises new and profound issues of national sovereignty. If one state becomes so overwhelmingly powerful on a global scale, in what sense do other states retain their full measure of sovereignty?

12.5 A mind experiment

We ought to take the rhetoric of American spacepower partisans with a very heavy dose of salt. Many of the schemes they speak of, particularly when they get to discussions of 'force application', the capability to attack terrestrial targets from space, are so unimaginably expensive and technically fantastic as to be doomed to fall of their own weight. A rule of thumb: if a given military task can be done by a terrestrial system, go with it. Basing observation, warning, communication, meteorological and navigation hardware in space has obvious advantages; otherwise, space is a difficult and pricey territory and ought to be avoided.

In crafting their vision statements and proposed doctrines, today's space warriors are inclined to cast aside cost considerations. They are 'big-picture' people. The idea that the US has the right to assume unilateral control of space is widely accepted in some Air Force circles, in the inner rings of the Pentagon, in certain think tanks, and perhaps within the Oval Office. We are the good guys, goes the argument. Why should anyone worry about US control of space? A mind experiment may offer a clue.

Imagine for a moment that another state had produced documents outlining why and how it could unilaterally achieve control of space (the Joint Chiefs of Staff and Air Force Space Command, the Pentagon's 'executive agent for space', have recently crafted such documents). Suppose that China or Russia had declared an intention to achieve full spectrum dominance in the military sphere, including space, by the year 2020. Assume further that the chief of Russia's or China's uniformed military services had said that 'our military is built to dominate all phases and mediums of combat. We must acknowledge that our way of war requires superiority in all mediums of conflict, including space. Thus, we must plan for and execute to win space superiority.' This was said by Richard B. Myers while he headed US Space Command (AW&ST, 1 January 2000). Myers went on to become chairman of the Joint Chiefs of Staff during the presidency of George W. Bush.

Alternatively, pretend that Iran, Syria or North Korea had told the world that it would build the capability to dominate 'the space dimension of military operations'. Or what if Britain, France, Germany or Japan had announced that it aimed to control space so that it could, if required, 'deny others the use of space'. What if Switzerland or Sweden, Austria or Australia, India or Indonesia had authored *Vision for 2020*, complete with a full-colour illustration of a space-based laser blasting a terrestrial target?

Choose your country – friend or foe, assertive or passive, kingdom, democracy, or dictatorship. Imagine further that the nominated country actually *had* the scientific, technical and financial resources to achieve its goals. That last requirement is a stretch, to be sure. The USA is the only country that can even aspire to space control. We are suspending disbelief here. Make a choice. Would you be surprised if Russia or China or any other country announced that it planned to control space in 20 years? Worried? Alarmed? Angry?

Rhetorical questions all. What right would *any* country have to unilaterally develop the capability to control space and to 'deny' access to others, if it so chose? If Britain, France or Japan had such plans, Americans would demand that Washington lean on the offending nation as hard as needed to force a recantation. If space-control rhetoric had come from India or Indonesia, the USA would call for condemnation in international fora and the imposition of draconian economic sanctions.

If, however, such measures failed, the world would have a new space race. Military dominance of near-earth space rather than putting men and women on the Moon would be the goal. The new space race would be hugely expensive; it would suck intellectual resources and scarce capital into black holes of mutual suspicion. It would compromise the ability of nations to meet everyday human needs. Worse, it would make fruitful cooperation on a host of pressing global problems less likely.

Nonetheless, let the race begin. The USA could not and would not let Country X or Nation Y take control of space. Reasonable people in Boston or Chicago or Seattle do not fret over Russian or Chinese satellites sliding overhead, unseen and unheard. That has been going on for decades. But ground-based lasers capable of disabling US satellites? That would be intolerable. Direct-ascent or space-based weapons capable of knocking out US satellites? Unacceptable.

What if, several years down the line, Country X or Nation Y actually developed workable space-based weapons. Kinetic-energy weapons capable of taking out the White House, the Capitol, the Statue of Liberty; unmanned spacebombers that could swoop down on the Pentagon without warning; orbiting lasers that could zap *Air Force One* as it wings across the Rockies. The prospect would be so horrifying as to require immediate action.

The USA may have the best of intentions. It may have no notion of ever denying access to space to another country except *in extremis*. It may have no wish to vaporize the satellites of other nations or to demolish buildings with kinetic-energy rods shot from space unless a war was in progress. It may not plan to ever shoot down planes with laser beams unless it was first attacked.

But what nation could afford to rely on the everlasting good intentions of another nation, even the USA?

12.6 Velvet glove, steel fist

US space warriors argue that the control of space involves nothing more sinister than building a navy to control the seas or an air service to command the air. The analogy is faulty. US air and seapower, while overwhelming, cannot be deployed everywhere at once. In contrast, space weapons, if developed, would be an 'always' thing, a pervasive Sword of Damocles, war machines orbiting overhead seven days a week, 24 hours a day, in times of either peace or war.

We have stepped through the Looking Glass and entered a Wonderland in which words and phrases mean whatever the space warriors want them to mean. US spacepower plans should not worry anyone, they say. US intentions are, and would remain, 'non-aggressive', a deterrent to bad actors and a threat to no one else (for instance, Rumsfeld, 2001, Section III; C 1: 'Impact on the Military Use of Space').

Everett C. Dolman is a professor at the Air Force School of Advanced Airpower Studies, the birthplace and nursery for airpower and now space-power doctrine since the 1920s. He argues in *Astropolitik* (2002) that the USA should 'endeavour at once to seize military control of low-earth orbit'. Only America can be depended upon to regulate space for the benefit of all.

'The military control of low-earth orbit would be for all practical purposes a police blockade of all [the world's] spaceports, monitoring and controlling all traffic both in and out... In time, US control of low-earth orbit could be viewed [by the rest of the world] as a global asset and a public good' (Dolman, 2002, pp. 157–9).

The assumption that other nations would be comfortable with a formulation even remotely like that is bizarre. How many nations could afford to be as generous in interpreting US intentions? What nation would be willing to play 'Mother-May-I?' with the USA in regard to its own national security? What other nation would be willing to be subject to changing US whims and geopolitical aims? By definition, a nation that can successfully control space during conflict would be able to deny access to space to any nation at any time. US control of space seems sensible and necessary to America's space warriors, but to other states, US control of space is more likely to suggest a velvet-glove hegemony that could turn to steel.

The National Security Strategy of the USA, issued in September 2002, offers little comfort to the already wary. Although the policy declares that it favours 'pre-emption', hitting bad guys before they hit us, it is a policy that also justifies *preventive* war. That is not mere semantics. In international law and practice there is a huge difference between 'pre-emptive' war and 'preventive' war. The former is often justified; the latter is unworthy of a nation that regards itself as law-abiding. One need not be a brain surgeon, or for that matter a rocket scientist, to understand how control of space would enhance a policy of full spectrum dominance and preventive war.

12.7   Unintended consequences

An attempt by the USA to achieve unilateral control of space and to place weapons in space would not be America's finest hour. It would be an insult to everyone on the planet. The Outer Space Treaty of 1967 treats space as a weapons-free sanctuary, barring nuclear weapons and 'other weapons of mass destruction' from space. However, there is a loophole in the treaty that US space warriors hope to exploit. In the 1960s treaty negotiators simply did not imagine that any nation would ever desire to place 'conventional' weapons in space – that is, weapons with precision effects. The USA, and only the USA, talks of doing just that. The USA, and only the USA, talks of developing and deploying a comprehensive space-control capability.

Space control, at face value, is not a bad idea. Control of space is necessary if humankind is to work its way towards a more humane future. But this aim should be pursued by international compact with vigorous enforcement provisions. Such a compact would be achievable and verifiable. Whereas work on some kinds of weapons systems, especially biological weapons, can be easily disguised, advanced work on anti-satellite weapons is not easily hidden. At some point, development must be carried out in the open, which means in space.

Given that visibility, it should not be impossibly difficult to design reliable verification procedures and technologies that would prevent either an arms race in space or a ground-based ASAT race. So far, however, the USA has given no indication that it will proceed in that direction. A treaty to prevent an arms race in space is not on the cards, at least as far as the USA is concerned. Just as surely as the matter is brought up each year at the Conference on Disarmament in Geneva, the USA blocks substantive work on such a treaty.

In September 2000, for instance, Robert T. Grey, then the US ambassador to the Geneva conference, offered this classic observation: 'The US agrees that it is appropriate to keep this topic under review. On the other hand, we have repeatedly pointed out that there is no arms race in outer space – nor any prospect of an arms race in outer space, for as far down the road as anyone can see' (Grey, 2000).

He was right, of course, at least regarding the first part of his assertion. At the time he spoke, there was no arms race in space because there was only one possible runner, the USA.

However, evaluating the wisdom behind a possible attempt by the USA to achieve control of space in the twenty-first century is not straightforward. Space-control enthusiasts are surely right when they say that America's vital interests must be vigorously defended. They are correct when they say that the USA, more than any other nation, relies on its space assets – military and commercial – to help it fight and to keep its economy vibrant.

However, space warriors are mistaken when they say that the USA must unilaterally achieve control of space to ensure its security. In a world based on the principle that nations are sovereign, unilateral control of space and weapons in space would raise profoundly troubling questions regarding national sovereignty. Most likely, it would be regarded by some states as an intolerable violation of global norms.

Many nations already hate or fear the USA, in part because of its techno-logical dominance. One suspects, though, that most states have already come to terms with the fact that, for the foreseeable future, the USA will continue to be the most powerful state the world has ever known, militarily, economically, and culturally. Is there a tipping point, a bridge too far, a line beyond which even a nation as benign as the USA cannot go without provoking some sort of reaction?

US spacepower partisans define space control as having the capability to grant access to space to the good guys and to deny access to the bad guys. Spacepower advocates frame that capability in the language of deterrence – a latent power that would be exercised *only* when necessary. They simply ignore the logical political consequences of that power. The USA could become the judge and jury regarding access to space by other nations. Why would any nation, even a friend, be content with that?

If the USA chooses to pursue a programme of space control, *and that choice had not been definitively made as this was written in 2004*, all bets would be off. The consequences would be unpredictable. Many nations would presumably go along with it, either because they are old friends and allies or because they are too poor to have any choice.

At least a few states would develop 'asymmetrical responses' to counter-balance the increase in spacepower. There is already considerable evidence that America's high-tech military lead is inspiring such strategies and programmes (for instance, the wealth of analyses available from the Defence Threat Reduction Agency at www.dtra.mil.). Low-tech nuclear weapons probably head the list, and they can be very low-tech indeed if delivered by truck or van instead of missiles. Biological and radiological weapons may not be far behind. Of course, there are endless possibilities for no-return-address terrorism, not to mention various forms of cyberwar, as others in this volume have pointed out.

The US pursuit of space control would be a new wild card in the continuing global poker game of power. The impact on other nations would be unpredictable, but surely some states that had been previously sitting on the fence would be so alarmed that they would take action.

Many hard-headed observers of the military scene regard space-control schemes as profoundly illogical and the possible deployment of actual weapons in space as unnecessarily provocative and hopelessly expensive. A devastating critique of space control and weapons in space was recently published in *International Security*, a highly respected journal. The principal authors were Richard L. Garwin, one of the world's leading physicists and a science adviser to presidents from Eisenhower to Carter, and Bruce M. DeBlois, a retired air force colonel who compiled and edited *Beyond the Paths of Heaven: the Emergence of Space Power Thought*, a basic spacepower primer published by the Air Force School of Advanced Airpower Studies.

Their article, 'Space Weapons: Crossing the US Rubicon', analysed the technical difficulties and extraordinary costs of the space-warrior schemes and found that scarcely anything on the space-warrior wish lists made any sense. In every instance, terrestrial systems could accomplish the military mission better – and far more cheaply. In conclusion, the authors said the best way to ensure US space security over the long haul was to promote the *non-weaponization* of space.

An aggressive [US] campaign to prevent the deployment of weapons by other nations might best be implemented as a US commitment not to be the first to deploy or test space weapons or to further test destructive anti-satellite weapons. A unilateral US declaration should be supported by a US initiative to codify such a rule, first by parallel unilateral declarations and then perhaps by the use of sanctions or force against actions that would imperil the satellites of any state. (DeBlois and Garwin, 2004, p. 84)

Knowledgeable men and women, including Garwin and DeBlois, have been saying such things for years. (In the case of Garwin, for two decades.) Nonetheless, the space-control train rolls on. High-level space warriors regularly assert that the US must pre-emptively develop the capability to control space because if America does not, another state will jump in. Peter B. Teets, undersecretary of the Air Force, put it this way in 2003:

> If we do not pursue control of space, then someone else will. If we do not exploit space to the fullest advantage across every conceivable mode of war fighting, then someone else will – and we allow this at our own peril. If we do not develop a new culture of space professionals – a new form of warfighter – then someone else may do so first, with dire consequences...Our success at wielding airpower has come with a realization that we need to do it before – and better than – anybody else. Let us do the same for space. (Tees, 2003)

12.8 'Last, best hope'

US control of space, says Professor Dolman, would place 'as guardian of space the most benign state that has ever attempted hegemony over the greater part of the world'. It would be a bold and decisive step and, 'at least from the hegemon's point of view, morally just' (Dolman, 2002, p. 158). The operative phrase here is 'from the hegemon's point of view'. It is likely that a number of other states would see it very differently.

If the USA still has moral authority in much of the world, it is because many hundreds of millions of people in other lands understand that the USA, despite its flaws, strives to be a reasonably fair, just and democratic society. Assuming unilateral control of space would not square with that perception.

In his second annual message to Congress in 1862, in the midst of Civil War, Abraham Lincoln spoke of the meaning of liberty and the symbolic importance of the USA to the world. The outcome of that conflict, he said, would determine whether the American people 'shall nobly save, or meanly lose, the last, best hope of earth'.

The belief in American exceptionalism, so nicely highlighted by Lincoln, has been both virtue and vice. It has helped make the USA a great and dynamic nation, but it has also caused the USA a lot of trouble over the years. The Vietnam War, in which more than 58,000 Americans and more than a million Vietnamese died, is testimony to how the USA can get things terribly wrong.

Yet the *idea* of America remains grand in conception, if not always in execution. The USA is the most open, the freest, and the most diverse society in the history of the world. The economic, political, cultural, and military power of the USA is enormous. Such power must not be misused.

An attempt by the USA to take unilateral control of space, to assume the role of global space cop, as Wernher von Braun first suggested, would be the ultimate expression of neo-imperial arrogance.

The next few years are critical. The USA can help repair its moral authority in the world by working with other nations to craft a treaty to prevent an arms race in outer space. Or it can lose that moral authority if it chooses to take unilateral control of space and to place weapons in space, weapons that would orbit above the heads of everyone, not just its enemies.

References

AW&ST, *Aviation Week and Space Technology*. Accessed online at www.awstonline.com.

Clinton, President W.J., *The Second Inaugural Address of President William J. Clinton*. Available at http://www.law.ou.edu/hist/clinton2.html.

Crane, C.C. *Bombs, Cities, and Civilians* (Lawrence, KS: University Press of Kansas, 1993).

DeBlois, B.M. *Beyond the Paths of Heaven* (Maxwell Air Force Base, AL: Air University Press, 1999).

DeBlois, B.M. R.L. Garwin, R.S. Kemp and J.C. Marwell, 'Space Weapons: Crossing the US Rubicon', *International Security*, 29(2) (2004), 50–84.

Dolman, E.C. *Astropolitik: Classical Geopolitics in the Space Age* (London: Frank Cass, 2002).

Emme, E.M. *The Impact of Air Power* (Princeton, NJ: D. Van Nostrand, 1959).

Futrell, R.F. *Ideas, Concepts, Doctrine: Basic Thinking in the United States Air Force, 1907–1960*, vol. 1 (Maxwell Air Force Base, AL: Air University Press, 1989).

Garrett, S.A. *Ethics and Airpower in World War II* (New York: St Martin's Press, 1993).

Grey, R.T. *U.S. Mission Geneva. Press Releases 2000*, (2000). Available at www.us-mission.ch/press2000/0915Grey.htm.

Harris, A. *Bomber Offensive* (London: Greenhill Books, 1990). The 1990 edition is a facsimile of the original, published in 1947, with a new introduction by RAF historian Denis Richards.

Irving, D. *Apocalypse 1945: the Destruction of Dresden* (Cranbrook, WA: Veritas Publishing, 1995).

JCS, *Joint Publication 3–14 Joint Doctrine for Space Operations*, 9 August 2002. Available at www.dtic.mil/doctrine/jpoperationsseriespubs.htm. For other documents, including the *National Security Strategy of the United States of America*, as well as *Joint Vision 2020* (which explores the role of information dominance and space assets), go to: www.dtic.mil/jcs.

LeMay, C.E. with MacKinlay Kantor, *Mission with LeMay: My Story* (Garden City, NY: Doubleday, 1965).

Lord, L. *Air Force Space Command at 21*, 19 September 2003. Available at http://www.peterson.af.mil/hqafspc/News/News-Asp/nws-tmp.asp?/storyid=03-180.

Markusen, E. and David Kopf, *The Holocaust and Strategic Bombing* (Boulder, CO: Westview Press, 1995).

McDougall, W.A. *The Heavens and the Earth: A Political History of the Space Age* (New York: Basic Books, 1985). The role of President Eisenhower in establishing the 'space-for-peaceful-purposes' policy is widely documented. However, McDougall's book is perhaps *the* basic text of the Space Age.

NYT. Available at www.nytimes.com/2002/04/09/opinion/09GART.html.

Rumsfeld, D. et al., *Commission to Assess United States National Security Space Management and Organization*, 2001. Paper copies of the 'Space Commission' report are exceedingly rare, but it is widely available online, for instance, www.space.gov.

Ryan, C. (ed.), *Across the Space Frontier* (New York: Viking Press, 1952).

M. Sherry, *The Rise of American Air Power: the Creation of Armageddon* (New Haven, CT: Yale University Press, 1987). (The previous five books cited explore the dark side of the Anglo-American strategic bombing campaigns in World War II. It is not possible to properly evaluate America's new way of precision war, based in large measure on information dominance and assets in space, without a clear understanding of that history.)

SpaceCom. Most, if not all, US Space Command documents are no longer easily accessible online. However, extensive commentary on *Vision for 2020* as well as the *Long-Range Plan* remains. A particularly helpful search engine is www.SearchMil.com.

Teets, P.B. 'Developing Space Power: Building on the Airpower Legacy', *Air and Space Power Journal*, 17(1) (Spring 2003), 11–15.

White, General T.D., Air Force chief of Staff, *Control of Space* (1955). Available at http://www.fas.org/spp/military/docops/usspac/lrp/ch05a.htm.

13

Threat Assessment and Protective Measures: Extending the Asia–Europe Meeting IV Conclusions on Fighting International Terrorism and Other Instruments to Cyber Terrorism*

Massimo Mauro

13.1 Introduction

This chapter provides a description of the Asia–Europe Meeting (ASEM) dialogue and cooperation framework, and a definition of cyber terrorism, detailing its current impact. In addition, some classes of cyber threats, protective measures, and possible policy options to achieve cooperation against cyber terrorism and other cyber threats in the Asian region are outlined.

13.2 The Asia–Europe Meeting (ASEM) framework

The ASEM is an informal process of dialogue and cooperation bringing together the 15 European Union member states and the European Commission with ten Asian countries, including Japan and Vietnam. The ASEM dialogue addresses political, economic and cultural issues, with the objective of strengthening the relationship between the two regions in the spirit of mutual respect and equal partnership.

The second ASEM seminar on e-commerce, held on 23 September 2002 in Helsinki, Finland, concluded that cyber security was a key priority area. Recommendations were issued to policy makers following the second ASEM

---

* The opinions expressed here are those of the author and should not be construed as representing the opinion of the Council of the European Union, of its General Secretariat, nor of any other European institutions.

Trade Facilitation Action Plan (TFAP) meeting on e-commerce, held on 24 September 2002 in Helsinki, Finland, these recommendations included:

II. Cyber Security

In compliance with the General Principles, ASEM partner countries should seek to achieve, through an appropriate mix of public and private sector action:

Compliance with internationally accepted best practices in the field of information and network security, such as the OECD [Organization for Economic Cooperation and Development] Guidelines for the Security of Information Systems and Networks;

1. Appropriate protection of critical information infrastructure;
2. An appropriate balance between, on the one hand, the confidentiality needs of businesses and consumers, and on the other hand law enforcement and national security objectives, in particular by respecting the principles set out in international best practices such as the OECD Cryptography Policy Guidelines, and with reference, where appropriate, to the Council of Europe Convention on Cybercrime;
3. Widespread availability of appropriate authentication and signing mechanisms to promote secure open electronic trade and commerce, in particular with reference to the UNCITRAL Model Law on Electronic Signatures.

According to the Cooperation Programme on Fighting International Terrorism, agreed at the ASEM IV meeting (22–4 September 2002, Copenhagen, Denmark), an ASEM seminar on anti-terrorism would be held in 2003, before the fifth ASEM Foreign Ministers' Meeting in Indonesia. This would be held in order to discuss how to strengthen the UN's leading role and ASEM cooperation on anti terrorism. However, no specific reference to cyber terrorism has, so far, been made.

13.3 Cyber terrorism: an urban legend?

Dorothy Denning (2000) defined cyber terrorism as 'unlawful attacks and threats of attack against computers, networks, and the information stored therein when done to intimidate or coerce a government or its people in furtherance of political or social objectives'.[1] The Stanford Draft Treaty provides a slightly different definition: 'cyber terrorism means intentional use or threat of use, without legally recognised authority, of violence, disruption or interference against cyber systems, when it is likely that such use would result in death or injury of a person or persons, substantial damage to physical property, civil disorder, or significant economic harm'.[2]

Although the general press and some government agencies have claimed that various terrorist groups have launched or are preparing to launch cyber attacks, no examples or qualified evidence confirming these claims, other than anecdotal accounts, have ever been provided. Cyber attacks, other than viruses or worms, detected so far seem to point instead at three main classes of originators:

1. Script kiddies:[3] the lowest form of cracker; script kiddies do mischief with scripts and programs written by others, often without understanding what they have done.
2. Financial criminals: unlike the previous group, far from trying to attract notoriety, this group attempts to penetrate financial systems with stealth, hoping to get some immediate personal financial benefit. The extent of the attacks originated by this group is very difficult to assess, as financial organizations (for example, banks) are notoriously coy about documenting such attacks. Most of these attacks have been proven to originate inside the organization being attacked.
3. Political opponents: many cases have typically occurred at the height of a political crisis or international conflict. Websites of a specific country, or of a specific organization, will either be defaced or submitted to intense DDoS (Distributed Denial of Service) attacks, preventing users from contacting them.

None of the above groups fully epitomizes the definitions of cyber terrorism that we have considered so far. A provisional conclusion is that there appears, at the time of writing, to be no imminent cyber terrorist threat to be reckoned with.

The Internet is a new thing, and new things can appear more frightening than they really are. Much of the early analysis of cyber-threats and cyber security appears to have 'The Sky is Falling' as its theme. The sky is not falling, and cyber weapons seem to be of limited value in attacking national power or intimidating citizens. The examples presented in this paper suggest that nations are more robust and resilient than the early theories of cyber terror assumed. To understand the vulnerability of critical infrastructures to cyber attack, we would need for each target infrastructure a much more detailed assessment of redundancy, normal rates of failure and response, the degree to which critical functions are accessible from public networks and the level of human control, monitoring and intervention in critical operations. This initial assessment suggests that infrastructures in large industrial countries are resistant to cyber attack.[4]

13.4   A taxonomy of real cyber threats

Let us attempt to establish a taxonomy of real threats. A tentative taxonomy of threats may be:

- Threats to the Internet infrastructure.
- Threats to individual networks or servers.
- Threats to critical infrastructures.

For design purposes, the Internet infrastructure is fairly resilient. Even an attack on DNS (Domain Name System) root server in October 2002,[5] which took the form of a relatively unsophisticated DDoS ping flood, was hardly noticed by general Internet users. Deployment of IPv6 (Internet Protocol version 6), which is supported and encouraged by the European Commission,[6] may assist with some security features unavailable in IPv4, although only a widescale deployment may allow us to reap these benefits. Router attacks are made possible by the inherent weaknesses of the routers themselves and may be corrected by the vendor, once detected.

Individual networks are sensitive to attacks according to the level of their defences. The focus keeps shifting: attacks which were previously of rather limited practical application, like ARP (Address Resolution Protocol) cache poisoning,[7] a form of 'man-in-the-middle' attack, which can only be used at subnet level and not across a WAN (Wide-Area Networks), is currently enjoying immense popularity because of the extraordinary growth in wireless networks and the multiplication of wireless access points, usually poorly secured (WEP [Wired Equivalent Privacy], or worse).

Attacks relying upon hostile vectors (worms, viruses, and so on) are nowadays probably one of the top priorities of individual network managers (be they corporate or governmental).

Critical infrastructure is a semantically ambiguous expression because of its context-dependent nature. In the USA, according to the Clinton administration:

> Critical infrastructures are those physical and cyber-based systems essential to the minimum operations of the economy and government. They include, but are not limited to, telecommunications, energy, banking and finance, transportation, water systems and emergency services, both governmental and private.[8]

One will note that the term 'cyber based' is rather vague, as is the expression 'minimum operations'. The Bush administration offer a definition of detailed critical infrastructure as follows:

> The information technology revolution has changed the way business is transacted, government operates, and national defense is conducted.

Those three functions now depend on an interdependent network of critical information infrastructures. The protection program authorized by this order shall consist of continuous efforts to secure information systems for critical infrastructure, including emergency preparedness communications, and the physical assets that support such systems. Protection of these systems is essential to the telecommunications, energy, financial services, manufacturing, water, transportation, health care, and emergency services sectors.[9]

The policy shift towards the inclusion of defence systems, in the aftermath of the September 11 attacks, is visible, even though it is obvious that strategically important US defence systems are not connected to the Internet, instead using EMP (Electromagnetic Pulse)-hardened dedicated networks, especially now that EMP or HPM (High-Power Microwave) weapons are a reality[10] independent of nuclear weapons.

In Europe there is a different focus: 'Networks and communication systems have become a key factor in economic and social development and their availability and integrity is crucial to essential infrastructures, as well as to most public and private services and the economy as a whole'. In addition, 'The security of transactions and data has become essential for the supply of electronic services, including e-commerce and online public services, and low confidence in security could slow down the widespread introduction of such services.'[11] In other words, the stress in Europe is on protecting network and communication services and e-commerce or e-government services.

In addition, due to the different historical development paths followed by the Internet in Europe and in the USA, the protection of the Internet infrastructure is given a different priority. In the USA, the existence of the National Science Foundation (NSF) backbone, predating the current Internet, allowed a systematic and logical backbone development and its subsequent adaptation to the ceaseless traffic growth. The existence of a robust Internet backbone is thus an axiomatic truth for Americans.

Such a backbone never existed in the anarchic Internet development in Europe and, hence, every service provider had to deploy his own network, relying on bilateral peering agreements with other ISPs (Internet Service Providers) to exchange traffic. This has led to the development of a patchwork structure of high-volume traffic nodes, which are extremely difficult to monitor and protect. The lack of a backbone structure, and the frequent dependency on US nodes to route even intra-European traffic, means that Europeans are more aware of the need to establish a more robust Internet.

However, since CERT (Computer Emergency Response Team) began to record incidents, there has not been to date (2004) a single case of attack against the Internet infrastructure itself which caused significant disruption. Other attacks, like the Nimda worm, had collateral effects on the Internet because the high scanning rate of the Nimda worm caused bandwidth denial-of-service conditions on networks with many infected machines.

13.5 Advanced defensive methods and different regional priorities

In the USA, much attention has been devoted to survivability studies. One recent definition of survivability is: 'the capability of a system to fulfil its mission, in a timely manner, in the presence of attacks, failures, or accidents'.[12] Survivability techniques usually rely upon methodologies for development and architectural policies, and have a broad scope. Their main objective is to allow operating systems which degrade gracefully under attack.

Biologically inspired methods against intrusion and propagation of hostile vectors seem to open promising avenues both in the USA and in Europe. Immunological intrusion detection models (inspired by vaccination techniques) have been extensively researched[13] and appear to be more effective than conventional hard-wired techniques.

Infection transmission reduction methods (similar to infection prevention methods in biological contamination) have also been designed (for example, throttling[14]) and proven to be effective in reducing propagation of viruses and of some types of worms. Against polymorphic[15] viruses, which are currently the hardest to detect, the combination of *in-vitro* study (in an isolated virtual machine, analogous to letting bacteria grow in a Petri dish) and of heuristics seems to be reasonably effective.[16]

In the USA there is a significant emphasis on methods which might be used in any environment to defend what they believe are critical infrastructure systems against any type of threat,[17] and if necessary to let them degrade gracefully. Alternatively, in Europe the stress seems to be first on maintaining the very infrastructure of the Internet, and then on keeping general web servers, whether they are used for electronic commerce or to provide government services, both secure and working.

Asian countries, with some regional tensions and a growing Internet presence, are likely to be more interested in defending information and communication infrastructures.

13.6 International and regional cooperation against cyber terrorism

In March 2004 the European Union set up a 'Network and Information Security Agency'[18] with various tasks:

(a) collect and analyse data, including information on current and emerging risks and, in particular, those which would impact on the resilience of critical communications networks and the information accessed and transmitted through them;

(b) provide assistance and deliver opinions within its objectives to the Commission and other competent bodies;

(c) enhance cooperation between different actors operating in the field of network and information security, inter alia by establishing a network for national and Community bodies;

(d) contribute to the availability of rapid, objective and comprehensive information on network and information security issues for all users by, inter alia, promoting exchanges of best practice on methods of alerting users, including those related to computer attack alert systems, and seeking synergy between public and private sector initiatives;

(e) assist, when called upon, the Commission and national regulatory authorities in analysing the implementation of network and information security requirements for operators and service providers, including requirements on data protection, that are contained in Community legislation;

(f) contribute to the assessment of standards on network and information security;

(g) promote risk assessment activities and encourage interoperable risk management solutions within organisations;

(h) contribute to the Community approach on cooperation with third countries including facilitating contacts with international fora;

(i) undertake any other task assigned to it by the Commission within its objectives.[19]

While some of these tasks, for example (b), (h) and (i), are probably not relevant to the discussion in hand, the other tasks could be entrusted to a similar Asian organization, yet to be formed. These tasks would be a practical implementation of the second ASEM TFAP meeting on e-commerce recommendations, and could contribute to increasing trust among collaborating countries, with a view to attaining a common goal.

It may also be worth noting that the legal framework provided by instruments such as the Council of Europe Convention[20] on cyber crime is, as its Article 5 shows, insufficient to deal with cyber terrorism (if and when it will occur) or with other threats to the information and communication infrastructure. A new multilateral treaty (such as the Stanford Draft Treaty) would be required to provide a comprehensive legal framework for mutual assistance in matters related to cyber threats and to ensure the possibility for the authorities of a country to act legally in pursuit.

13.7 Concluding statement

Cyber terrorism is, for the time being, only a theoretical possibility, and there is no current need to rush for defensive weapons. Methods to defend from other real cyber threats exist and are continuously being developed.

Some models for international cooperation in fighting cyber threats may be adapted to the diverse realities of different regions of the world.

Notes

1. D. Denning, 23 May 2000. Available at http://www.terrorism.com/documents/denning-testimony.shtml.
2. A.D. Sofaer, S.E. Goodman, M.F. Cuéllar, E.A. Drozdova et al., 'A Proposal for an International Convention on Cyber Crime and Terrorism', *The Information Warfare Site* (2000). Available at http://www.iwar.org.uk/law/resources/cybercrime/stanford/cisac-draft.htm.
3. *The New Hackers Dictionary*. Available at http://www.hack.gr/jargon/html/S/script-kiddies.html.
4. J.A. Lewis, 'Assessing the Risks of Cyber Terrorism, Cyber War and Other Cyber Threats', *Centre for Strategic and International Studies*, December 2002. Available at http://www.csis.org/tech/0211-lewis.pdf.
5. R. Naraine, 'Massive DDoS Attack Hit DNS Root Servers', *Enterprise*, 23 October 2002. Available at http://www.internetnews.com/ent-news/article.php/1486981.
6. European Commission, *European Commission IPv6 Task Force*. Available at http://www.ec.ipv6tf.org/in/i-index.php.
7. B. Fleck and J. Dimov, *Wireless Access Points and ARP Poisoning: Wireless Vulnerabilities that Expose the Wired Network* (2001). Available at http://www.cigitallabs.com/resources/papers/download/arppoison.pdf.
8. US Department of Homeland Security, 'White Paper. The Clinton Administration's Policy on Critical Infrastructure Protection: Presidential Decision Directive 63', *Information Analysis and Infrastructure Protection*. May 1998. Available at http://www.ciao.gov/publicaffairs/pdd63.html.
9. The White House, *Presidential Executive Order on Critical Infrastructure Protection*, 16 October 2001. Available at http://www.whitehouse.gov/news/releases/2001/10/20011016-12.html.
10. R.A. Kelis, 'The Radio Frequency Weapons Threat and Proliferation of Radio Frequency Weapons', *Statement before the US Congress Joint Economic Committee*, 25 February 1998. Available at http://www.house.gov/jec/hearings/02-25-8h.htm.
11. The Council of the European Union, 'Council Resolution of 28 January 2002', *Official Journal of the European Communities*, 28 January 2002. Available at http://europa.eu.int/eur-lex/pri/en/oj/dat/2002/c_043/c_04320020216en00020004.pdf
12. N.R. Mead, R.J. Ellison, R.C. Linger, T. Longstaff and J. McHugh, 'Survivable Network Analysis Method', *Carnegie Mellon Software Engineering Institute*, September 2000. Available at http://www.sei.cmu.edu/publications/documents/00.reports/00tr013/00tr013chap02.html
13. See, for example, Anchor et al., The University of Wales (2002). Available at http://www.aber.ac.uk/~icawww/Proceedings/paper-32/Anchor-ICAR1S-2002.pdf
14. M.M. Williamson, *Throttling Viruses: Restricting Propagation to Defeat Malicious Mobile Code*, 17 June 2002. Available at http://www.hpl.hp.com/techreports/2002/HPL-2002-172.pdf.
15. M. Landesman, *Antivirus Software*. Available at http://antivirus.about.com/library/glossary/bldef-poly.htm.

16. Symantec, 'Understanding and Managing Polymorphic Viruses', *The Symantec Enterprise Papers*, vol. XXX (1996). Available at http://www.symantec.com/ avcenter/reference/striker.pdf.

17. 'Our nation's information and telecommunications systems are directly connected to many other critical infrastructure sectors, including banking and finance, energy, and transportation. The consequences of an attack on our cyber infrastructure can cascade across many sectors, causing widespread disruption of essential services, damaging our economy, and imperilling public safety. The speed, virulence, and maliciousness of cyber attacks have increased dramatically in recent years. Accordingly, the Department of Homeland Security would place an especially high priority on protecting our cyber infrastructure from terrorist attack by unifying and focusing the key cyber security activities performed by the Critical Infrastructure Assurance Office (currently part of the Department of Commerce) and the National Infrastructure Protection Center (FBI).' The White House, 'Information Analysis and Infrastructure Protection', *Department of Homeland Security*, June 2002, http://www.whitehouse.gov/deptofhomeland/sect6.html.

18. See The European Network and Information Security Agency website, http:// www.enisa.eu.int.

19. The European Commission, 'Proposal for a Regulation of the European Parliament and of the Council. Establishing the European Network and Information Security Agency', *Commission of the European Communities*, 11 February 2003. Available at http:// europa.eu.int/information_society/eeurope/news_library/documents/nisa_en.pdf.

20. Council of European Convention, *Convention on Cybercrime*, 23 November 2001, http://conventions.coe.int/Treaty/en/Treaties/Html/185.htm.

14
Policy Laundering, and Other Policy Dynamics

Dr Gus Hosein

14.1 Introduction

Laws and policies are usually regarded as national affairs, for consideration in national policy discourses. This chapter argues, however, that we need to study the dynamics of modern policy development with an eye to international activities. I contend that this may involve focusing on three separate aspects of the policy process: *policy laundering, modelling,* and *forum shifting. Policy laundering* is a practice where policy makers make use of other jurisdictions to further their goals, and in so doing circumvent national deliberative processes. *Modelling* occurs when governments, overtly through calls of harmonization or subtly through quiet influence and translating of concepts, shape their laws based on laws developed in other jurisdictions. *Forum shifting* occurs when actors pursue rules in inter-governmental organizations (IGOs) that suit their purposes and interests, and when opposition and challenges arise, shift to other IGOs or agreement-structures. This chapter will review these dynamics and provide an example of their instantiation, being the initiatives on cyber crime emerging from the Group of Eight industrialized countries (G8) and the Council of Europe.

It is possible to perceive policy laundering as a natural extension of the power of governments. In the 1980s, and particularly in the 1990s, we saw a significant rise in the number of IGOs, and in the number of treaties emerging. This is, as 'they' say, the age of globalization, and global interdependence. It is inevitable that international action is required, through some form of global consensus. Norms and policies are pursued at these international bodies in order to establish international treaties and transform global norms. This benign view of power and international relations is complemented by a more cynical view that with the growth of IGOs, it is inevitable that governments will use these institutions to their advantage – just as they have used all other institutions in the past. Thus, the global norms and international treaties represent the interests of those with the greatest sway and influence, or at least those countries who sat at the table. Regardless of your view of power and benign interests, these international institutions are used as policy fora, and it is worth scrutinizing the dynamics involved in them.

There are two immediate implications of these new policy dynamics. First, national consultative processes disappear or are weakened, as important policy decisions take place outside our traditional democratic institutions. Frequent calls for 'harmonization' through treaty ratification and perceived international obligations inhibit the likelihood and effectiveness of traditional national deliberation, while these treaties are negotiated in closed environments.

A second implication is that policies are shaped by foreign interests and foreign processes. For instance, the European Union privacy practices have been under review because of the influence of recent US laws on travel documentation and procedures. Rules on trans-border data flows and the collection of fingerprints are now being renegotiated through international agreements based on laws created in another jurisdiction.

I am not among those who would say that this is an inevitable outcome of globalization and the balance of power and interests. Nation-states are losing power, some would argue, and, they continue, interdependence and international negotiation through multilateral organizations are ideal. My point is simultaneously both more cautious and more concerned: some interests of the state are being met, whilst deliberation is being relocated. States are not losing power; the harmonization and standardization of policies in the area of focus for this chapter involve proposals to *increase* the power of law enforcement agencies. The autonomy of the state is not being threatened; it is the capacity for democratic discourse that is overshadowed.

National governments are usually entitled to enact and enforce laws within their jurisdiction; it is, after all, their sovereign right to do so. There are conditions that arise where this sovereign right is questioned, or where conflicts arise. One such conflict arises when there is *trans-border activity*: where activity occurs beyond the jurisdiction that affects the ability of the sovereign jurisdiction to enforce its laws. This renders doubtful the viability of potential laws to solve this very problem. Enforceability becomes immediately questionable: activity may occur beyond the jurisdiction of national law, regulating national activities is fruitless and hazardous economically (Sun and Pelkmans 1998).

With data flows within transnational digital networks and the associated products and services, the trans-border problems are exacerbated. Action may occur at a distance, where the overflow of activity can occur without an individual having to physically enter the jurisdiction. Whether it is the penetration of computers or the downloading of pornography, this conduct can occur across borders, preventing law enforcement agencies – with their traditionally bordered jurisdictions – from conducting investigations and gathering evidence.

Establishing national policies amidst transnational data networks and flows is a pressing technology policy issue. It is not new, nor as infeasible as often presumed, or argued legally (cf. Johnson and Post, 1996). These presumptions and arguments have been responded to by legal theory and practice. Theoretically, it is now argued that the infeasibility arguments exaggerated the problems and promises of data flows (Goldsmith 1998, c.1130). In so doing, the arguments failed to recognise that there have been many multijurisdictional regulatory problems involving trans-national transactions in other fields of law (Goldsmith 1998, c.1200–01).

The debate has long moved on beyond the halls of academia, however. There are continuous calls for solutions to the challenges of trans-border data flows, trans-border criminal activity, and trans-border terrorist activity. The solutions that are preferred are not distinct from the seemingly national laws that have been established in response to national data policies (for example, data protection and privacy, electronic commerce), criminal policies (for example, investigatory powers and capacities), and anti-terrorism policies (for example, blanket surveillance, immigration procedures). Nor are the relationships between international and national solutions necessarily clear-cut. The next section considers some of these solutions.

14.2 At the international level: the Council of Europe and the G8

The activity of inter-governmental organizations appears, in many cases, to lead the way in developing policy in the 'age of globalization'. If our policy challenges are international in nature, and the infrastructure of trade and communications is also global, then, so the logic goes, we need global solutions developed by international fora. And these international fora are eager to be both active and relevant.

For a number of years, two international bodies were developing agreements for international cooperation for 'high-tech' or 'cyber crime'. Since 1995, the Group of Eight industrialized countries (G8) has been meeting regularly on a formal basis to discuss harmonizing methods, and creating new investigative powers and means of cooperation. Similarly, the Council of Europe (CoE), the 43-member-state international treaty-making body, has laboured to create the Convention on Cybercrime since 1997.

The work product of the CoE in cyber crime and the G8 on high-tech crime are worthy of deeper analysis, which is carried out elsewhere. Here a brief account of the story will have to suffice. The CoE convention on Cybercrime (ETS 185) consists of three components: a set of substantive crimes to be enshrined in law that includes hacking, child pornography, and copyright circumvention; a set of surveillance capacities that ratifying countries are expected to enable for use by their law enforcement authorities; and a regime for mutual legal assistance and extradition amongst ratifying countries.

This last component of the CoE convention is the most alarming: the creation of a broad mutual legal assistance agreement. Cooperation is particularly problematic as the convention tries to do away with traditional concerns for dual criminality; in fact, it dissuades and sometimes prevents countries from refusing assistance to another country on these grounds. The convention may create situations where a country will be required to collect evidence on an individual without any contravention of its own domestic law having occurred.

The convention was drafted by a group of representatives from national departments of justice and home affairs – most notably Canada, France, Germany, the United Kingdom, and the USA. It is worth noting that two of these countries (Canada and the USA) are merely observer nations to the CoE. As a result of the formulation and consultation processes, the convention represents the interests of law enforcement agencies, while all but ignoring privacy and civil liberties protections.

Drafted in relative secrecy in the period 1997–2000, a consultation process was opened in April 2000. Few changes were achieved in the consultation stage, however, despite the activities of representatives from industry and civil society. Both industry and civil society organizations expressed considerable opposition to the convention on a number of grounds, including the formulation process, invasiveness, costs and burdens, lack of due process provisions, and the presence of ambiguous language within the body of the convention.

The CoE responded to these appeals by making repeated promises that the opportunity for consultation and democratic participation would arise on a case-by-case basis at the national level at the time of signing and ratification.

At the same time as the convention was being negotiated, the G8 Lyon Group on Transnational Organised Crime was working on the problem of high-tech crime. Meetings began in 2000 to consult with industry in the G8 countries on proposed investigative powers. The conclusions of the first meeting, held in Paris in May 2000, were uncertain, as industry and governments diverged in their interests and statements. The final communiqué articulated some of these concerns, including civil liberties and privacy, maintaining the powers of law enforcement agencies, defining a clear and transparent regime to combat 'cyber criminality', and ensuring free and equitable market development to certify good conditions for industry, while evaluating the effectiveness and consequences of the policies (G8 Lyon Group, 2000).

In October 2000, the Berlin meeting ended with some further clarifications, but again agreed that further work was still required. Particular concerns were articulated regarding the quantification of costs of proposed measures, the implications of technology development, continuing law enforcement concerns about the availability of data, evolving definitions of the different types of service and service provider, process and procedures in mutual legal assistance, the integrity of the data, and privacy and civil liberties. The conclusions of Berlin set the stage for Tokyo in May 2001, as further research and consideration were required.

The Tokyo summit did not end on a high note; in fact, the continued inability to agree left the future of such industry–government dialogue in question. In a summary report of the Tokyo meeting by the Canadian Department of Foreign Affairs and International Trade, the complexity of the issues was placed as an obstacle to gaining any progress through the meetings: 'While deemed to be a success, the Tokyo Conference highlights certain deficiencies in the G8 government/industry dialogue process that will have to be addressed before future meetings are convened' (Purdy, 2001).

The earlier perception that further negotiations with industry were needed was corrected by the events of September 2001. Among articulations of 'greater urgency to this work' (G8 Lyon Group, 2001), draft G8 responses to the terrorist attacks on the USA were released in November 2001. These included calls for alterations to privacy laws to cater for 'public safety and other social values', permitting domestic law enforcement agencies to serve foreign preservation and real-time access instructions to domestic service providers after expedited approval, ensure expedited preservation and real-time access to traffic data, expedited mutual legal assistance even if there is no violation of the domestic law of the requested state, and to encourage 'user-level authentication' for appropriate uses. The earlier opposition had been erased from memory.

These recommendations were included as official documents at the Mont-Tremblant summit of Justice and Home Affairs ministers in May 2002. They included a call for governments to decide which information is useful for public safety purposes (G8 Justice and Interior Ministers, 2002), drawing directly from the Government–Industry Workshop documents detailing the types of law enforcement powers and procedures to be considered (Canadian Delegation, 2001; G8 Justice and Interior Ministers, 2002); and an official statement of how data protection and privacy regimes 'seriously hamper public safety' (G8 Justice and Interior Ministers, 2002). The summit leaders also addressed international treaties, including the Council of Europe convention.

14.3 At the national level

In this instance of high-tech crime policy, the dynamics are worth noting. The accusation of policy laundering was directed towards the USA on a number of occasions, with claims that the USA was pushing other countries to adopt the CoE convention even when it had no clear intention of doing so itself. The US ratification process is detailed and lengthy as it involves approval by the US Senate, in contrast to other countries whose executive arms of government have more sway over ratification procedures.

Some countries have already chosen to ratify the convention and use it as a model, including some that are not part of the CoE. At the time of writing, only Albania, Croatia, Estonia, Hungary, Lithuania, Romania, Slovenia, and Macdeonia have ratified, although there have been discussions in Bulgaria, Canada, France, Japan, and the convention has entered into the discourse in Australia. In July 2001 the Australian government announced that its bill on computer crime, which requires users to provide encryption keys, was to be based on the Convention (Attorney General, 2001). It is important to note that the convention contains no such requirement.

Japanese parliamentarians and civil society have relayed to the author their concerns that a national consultation on the ratification of the CoE convention is unlikely, under some pressure from foreign governments. At a US–Japan Information Systems & Network Security Forum in September 2003, the joint statement included the following paragraph:

> The United States and Japan also affirm the importance of multilateral cooperation for cybersecurity, including the international adoption of the Council of Europe Convention on Cybercrime. (United States and Japan, 2003)

and affirmed that:

> The Governments are encouraged to work within the appropriate multilateral fora – such as APEC, the G-8, and OECD – to implement cybersecurity and cybercrime recommendations and action plans that are adopted in these fora. (United States and Japan, 2003)

even as the national discourse on the convention is apparently taking place.

In Canada, after a number of years of considering other countries' policies on lawful interception, in 2002 the government finally began a consultation on 'lawful access'. The consultation document makes clear that much of the consultation is in response to the requirements of the CoE convention.

> The Convention calls for the criminalization of certain offences relating to computers, the adoption of procedural powers in order to investigate and prosecute cyber-crime, and the promotion of international cooperation through mutual legal assistance and extradition in a criminal realm that knows no borders. The Convention will help Canada and its partners fight crimes committed against the integrity, availability and confidentiality of computer systems and telecommunications networks and those criminal activities such as on-line fraud or the distribution of child pornography over the Internet that use such networks to commit traditional offences. (Department of Justice, Industry Canada, et al., 2002)

Interestingly, the document only goes on to discuss some of the surveillance capabilities and some criminalization; it fails to discuss international cooperation and mutual legal assistance. In a sense, the Canadian government used the CoE convention as the basis for increasing investigatory powers, but apparently it does not wish to ratify the convention and does not wish to adhere to cooperation requirements from other countries.

More recently, as the Council of Europe Convention on Cybercrime was introduced to the Senate Foreign Relations Committee, the chair of the committee stated immediately that, along with three other international treaties under consideration in that same hearing,

> I commend the U.S. officials who have worked on these agreements for negotiating documents that command wide support. Some of these agreements are the product of years of dedication and patient negotiations. Prompt ratification of these agreements will help the United States continue to play a leadership role in international law enforcement and will advance the security of Americans at home and abroad. (Lugar, 2004)

For what it is worth, there is extensive uncertainty regarding the Convention amongst both industry and nongovernmental organizations (NGOs), particularly regarding the lack of consultation in its development; and the US government is aware of this. This is what makes Senator Lugar's comments so defining of these dynamics: even during the negotiation process the USA had tried to be open (and was the only country to do so, to my knowledge) but then during the ratification process, just when the US procedure was most rigorous, the Chair of the Committee was calling for 'prompt' ratification and was bundling the convention with two other international agreements. International agreements and 'international leadership' is now the language of reduced national deliberation.

14.4   The international–national dance: traffic data retention

Data retention is an interesting example of the synergy between national and international policy making. The process began in the UK, to some extent, and ended up in the G8, the EU, and finally back in the UK. This route leads to many unforeseen consequences, however.

In August 2000, a number of UK law enforcement agencies submitted a proposal to the UK Home Office to require the retention of communications traffic data for up to seven years by a central government authority (Gaspar 2000). At that time this retention of sensitive information, including telephone traffic data, internet e-mail traffic data, and perhaps even website viewing data, faced significant resistance in the public discourse. In particular, this retention policy is contrary to data protection principles, in accordance with the EU Data Protection Directive 1995 (European Union

1995) as implemented UK law under the Data Protection Act 1998, that calls for the destruction of sensitive data when it is no longer required.

At around the same time, the EU was working on 'updating' its data protection directives to cater for electronic commerce and transactions. In this forum, the UK began calling for a change to the Data Protection Directive on Electronic Communications; a change that would *permit* member states to allow for data retention. There was some discussion of the issue, but little progress. There was consideration of including data retention in the CoE convention, but that was not accepted by the CoE. The G8 also discussed data retention, as was mentioned above, but industry originally rejected the proposals.

In October 2001 President George W. Bush wrote a letter to the president of the European Commission recommending changes in European policy, to '[c]onsider data protection issues in the context of law enforcement and counterterrorism imperatives', and as a result to '[r]evise draft privacy directives that call for mandatory destruction to permit the retention of critical data for a reasonable period' (Bush 2001). This was building on previously articulated concerns that '[d]ata protection procedures in the sharing of law enforcement information must be formulated in ways that do not undercut international cooperation' (United States Government, 2001).

Very similar language later appeared in the G8 documents from the May 2002 summit, even though it had previously been rejected by the G8 Government–Industry meetings, for instance, in the pre-September 11 era.

> Ensure data protection legislation, as implemented, takes into account public safety and other social values, in particular by allowing retention and preservation of data important for network security requirements or law enforcement investigations or prosecutions, and particularly with respect to the Internet and other emerging technologies. (G8 Justice and Interior Ministers, 2002)

Over these ensuing months, EU law was changed to allow for data retention. In December 2001 data retention was introduced and passed under the UK's anti-terrorism law in response to the events of September 11. A significant number of other member states have since passed similar retention laws.

Now there is consideration at the EU that the Directive did not go far enough. The 2002 Directive allowed for countries to pass retention laws. At the time of writing, however, the situation in Europe is fragmented: some countries allow retention, others do not. There is some pressure on the EU from the Justice and Home Affairs ministers to finalize a Framework Decision on data retention that would require all member states to have policies allowing for retention.

This forum shifting is alarming for a number of reasons that lie beyond the scope of this study. One particular concern, however, is that the US itself does not have a policy on data protection, nor, in turn, a policy on data retention; regardless, it asked the EU to change its privacy practices. Meanwhile, the UK failed with its original policy nationally and so shifted to the EU to allow the UK to pass such laws. Now that the EU has allowed for retention, there are calls for standardization and harmonization across Europe to ensure that other countries adopt similar laws. The confusing fact here is that policy laundering is indeed taking place, but it is hard to see who exactly is doing the laundry.

14.5 Democratic challenges and international opportunities

Sometimes it can be said that policies are homegrown, negotiated and deliberated needs of law enforcement agencies and other government departments. This is a simplistic view of policy, however; policies are being decided at international fora. It is clear now that domestic political actors need a clearer understanding of the international activities. To understand national policy dynamics it is therefore important to note the activities of inter-governmental organizations, and the conduct of national governments whilst negotiating agreements.

IGOs have been busy, particularly with regard to issues relating to digital and terrorism policies. But the vast majority of these meetings of inter-governmental organizations are closed to public review, participation, and consultation. In the coming months and years, however, their work, findings, conclusions, and conventions are likely to affect national policy discourse.

This effect is particularly likely in the case of the European Union, and primarily the output of the Council of the European Union, where the national governments alone have representation, where there is more of a binding requirement to enact policies at the national level.

14.5.1 Possible remedies

Greater attention must be paid to the activities of inter-governmental fora. Policies are being decided, and then adopted nationally, and this is being undertaken done selectively. These policies are not limited to mere cyber crime issues; for example, the UK is moving towards secure identity documents based on the discussions of the G8 and the International Civil Aviation Organization (ICAO); it has not engaged in a national discourse, but is deciding on using these passports as the grounds for payment of a national ID card. Other policies, such as transfer of passenger data to third countries, and even anti-terrorism policy, are being developed outside of national deliberative processes.

As countries move to ratify and implement policies agreed at inter-governmental organizations, the role of national NGOs is increasingly called into question. NGOs are for the most part focused on national policy developments, and are busy enough at that level. Now they also have to monitor the processes and outputs of IGOs that do not always operate openly. During the formulation of the cyber crime convention, the Council of Europe argued that consultation is ideally a national process, and not the responsibility of the CoE itself. While this may be true with respect to its current mandate, ratification may not be the ideal time to discuss serious problems with the convention through national policy discourses once there is already a perceived need to adjust national law accordingly. IGOs must change their mandates to include consultation, perhaps through requiring national consultation prior to the negotiation of charters, agreements, and treaties; otherwise, the sincerity of the political discourse is highly questionable.

Most importantly, civil society and other actors can interpret these instruments liberally. Generally speaking, these international documents are written in vague language or allow for caveats. While this vague language was set to meet the interests of the drafters and the adopting member states (to prevent overly restrictive language from preventing countries from convincing recalcitrant parliaments), the vagaries may be used to the advantage of lesser players. The CoE convention allows for a number of clarifications that can be complied with in a creative way. By reinterpreting international obligations, it may be possible to restrain government within what little national discourse there may be from increasing the powers of the state unnecessarily under the claim of necessity.

14.5.2 Flexibly interpreting and creative compliance

Despite the ambiguous definitions and terminology within the convention, in accordance with its Article 1, ratifying states are not obliged to copy the definitions directly. Many key concepts lack definition altogether. This is open to abuse – for example, the definition of 'traffic data' can be carried out in a technology-specific way in order to minimize the harm to civil liberties. Ideal definitions must acknowledge that some transactional data can be highly sensitive, including – but not limited to – location data, websites visited, computer names, chatroom conversation data, and search parameters. There may be a temptation to define traffic data for the Internet in a similar manner to traffic data for the telephone system; however, we would warn against such a weak distinction amongst technologies; telephone data is not the same as location data, for example. Once the sensitive data types are noted, lawful access policies of preservation, production, and real-time surveillance should vary according to the data required and produced. While interpretations may create legal inconsistencies across borders, even amongst countries with similar legal systems – for example Canada, the UK, and the USA – the definitions already vary greatly (Escudero-Pascual and Hosein 2003).

It is also possible to raise the bar, using interpretive flexibility. For example, the convention's Article 15 requires that countries ensure that the powers and procedures implemented in accordance with the convention are subject to conditions and safeguards in national law to provide for the adequate protection of human rights and liberties. This includes incorporating the principles of specificity and proportionality; ensuring adequate judicial supervision; assuring due process is followed through such measures as providing sufficient grounds prior to applying the powers; and limiting the scope and duration of these powers. We can therefore recommend that a ratifying state conduct an open dialogue of the invasiveness of all of these procedures and regarding the need to update and enhance human rights protections in the light of new technological developments, in accordance with the requirements of the convention. Different technological infrastructures involve different impacts on rights, responsibilities and the legitimate interests of third parties. The convention calls for countries to consider these issues in Article 15.3. Finally, the convention can be interpreted, liberally, as requiring judicial authorization for access to sensitive data. In some countries, notably the UK, this would be revolutionary as politicians and police agencies currently authorize such access.

Similarly, in the area of international cooperation, limitations on the requirements are quietly stated by the convention. Under Article 23, the convention appears to require the cooperation amongst states in the investigation of criminal offences and for the collection of evidence. There are, however, a number of grounds for refusal and reservations that a country may pursue. In the case of extradition under Article 24, a state may refuse to extradite an individual if it is not satisfied that all of the terms and conditions of the extradition are adequate. In accordance with Article 28, the requested party may restrict cooperation through the sharing of investigative data on the requirement that it not be used for investigations or proceedings other than those stated in the request. NGOs and other smaller players can perform a role in the negotiation of these terms.

Finally, a state may refuse requests for cooperation if it believes that the offence being investigated is a political offence or connected with a political offence, or prejudicial to its sovereignty, ordure public or other essential interests. Civil society may take this opportunity to nurture a national dialogue on what qualifies as a 'political offence', 'ordure public', and 'essential interests', and the procedures through which such qualifications are deliberated on a case-by-case basis. Particularly, it can easily be argued that any individual exercising their political rights should be protected by the 'political offence' clause.

Our goal must be to see strict definitions negotiated at the national level to take advantage of the ambiguities within the text of the convention; the minimization of the application of the measures of surveillance; clear processes of authorization and oversight to the use of invasive surveillance techniques; and the requirement of dual criminality prior to the use of any of these powers. If we do not do it, then law makers will, but in their own interests.

14.5.3 Centralized policy counter-moves

Another strategy is to engage these policies at their centralized point. While the CoE and the G8, and even the EU, are not recognised for the consultation processes, it is still possible to draw attention to their work. At the CoE, civil society actively wrote letters to the CoE Secretariat and gained media interest that would otherwise have been lacking. Within the G8 consultations, through the US government a number of NGO representatives managed to gain invitations to the summits; other countries can be forced to be as open as the US delegations were.

At the EU level it is possible to appeal to legal instruments. While not strictly a signatory to the European Court of Human Rights (ECHR), it is possible to argue that the EU's initiatives in data retention, while even in their draft modes, are legally problematic. That is, under a number of articles of the Convention on Human Rights, blanket data retention is illegal. A legal opinion drafted for use at the level of the EU may hit the policy centrally. Even if such a strategy centralized policy counter-move fails at the EU level, because many of the national laws are very similar to one another, and all EU states have to abide by the ECHR, a legal challenge can be interpreted and adapted and used locally.

14.6 Concluding remarks

With all of the changes in our political systems that have taken place over the past twenty years, our current predicament sounds so reasonable: globalization and international technologies and threats require global solutions. The rise of IGOs for policy deliberation and formation are, perhaps, inevitable outcomes.

This is a view that I would be inclined to reject, not because it is untrue, but, rather, because it is hazardous. Instead in this chapter I have highlighted that there are dynamics at play that are worthy of consideration in order to understand the darker side to 'international cooperation' and other such 'inevitabilities'. As these fora are used for policy deliberation and formation, democratic process and procedures are transformed. Multilateralism, therefore, is not a good in itself, but just another manifestation of power politics and the balancing of interests. Those with sway and influence in international fora, and those who are sitting at the negotiation tables at these meetings, have unprecedented national powers.

The policy landscape is thus transformed, and not in a favourable shape for national deliberation. Policy laundering is used to promote policies at international fora, only to bring them back home for 'ratification' or 'harmonization' with a minimal amount of debate. Both the USA and the UK can be accused of pushing ideas in international fora, and then are seen to pursue these policies actively at home, without claiming responsibility in the first place. Modelling is used to copy laws and international agreements and treaties, incorporating the language used abroad into national law without adequate consideration. Merely copying the language of 'traffic data' in the CoE Convention on Cybercrime may involve a radical reconsideration of national laws on due process and law enforcement powers, but this debate may never occur. Finally, forum shifting is used to pursue policies across IGOs until adequate homes are found. Data retention was pursued at the G8, and then pushed at the EU, where it found its natural home for EU member states who wished to find the path of least resistance in passing national laws.

Unless capacities are developed, these dynamics will diminish our capacity to act. Non-state actors need to develop a better repertoire of strategies in order to deal with these closed fora and their persuasive internationalist language. This form of activity is increasing, with every new treaty and international agreement, model law, recommendation and resolution emerging from all of these inter-governmental organizations.

Keeping track of all of these activities is not easy. The goal keeps on shifting, the policies keep on being transplanted and the calls for harmonization and international cooperation increase; and yet we do not appear to be following the match very closely. We need to pay attention to these dynamics in order to understand where the game is being played. Then we can create new structures of accountability for the players. There are opportunities; we may utilize some countermeasures. This will require a refocusing of our attention.

References

Attorney General *Cybercrime Bill 2001 Second Reading Speech by the Attorney General*, The Parliament of the Commonwealth of Australia, 2001.

Bush, G.W. 'Letter from President George W. Bush to Mr Romano Prodi, President, Commission of the European Communities. Brussels', forwarded by the Deputy Chief of Mission to the European Union, 16 October 2001.

Canadian Delegation *Discussion Paper for Workshop 1: Potential Consequences for Data Retention of Various Business Models Characterizing Internet Service Providers*, Tokyo, G8 Government–Industry Workshop on Safety and Security In Cyberspace, 2001.

Department of Justice, Industry Canada, et al. *Lawful Access – Consultation Document*. Ottawa, Government of Canada, 2002, p. 21.

Escudero-Pascual, A. and I. Hosein The Hazards of Technology-Neutral Policy: Questioning Lawful Access to Traffic Data. *Communications of the ACM*, 2003. Accepted for publication 24 October 2002.

European Union *Directive 95/46/EC of the European Parliament and the Council of 24 October 1995 on the protection of individuals with regard to the processing of personal data and on the free movement of such data*: 0031–0050 (1995).

G8 Justice and Interior Ministers *Data Preservation Checklists*. Mont-Tremblant, G8 Summit, 2002a.

G8 Justice and Interior Ministers *G8 Statement on Data Protection Regimes*. Mont-Tremblant, G8 Summit, 2002b.

G8 Justice and Interior Ministers *Principles on the Availability of Data Essential to Protecting Public Safety*. Mont-Tremblant, G8 Summit, 2002c.

G8 Justice and Interior Ministers *Recommendations for Tracing Networked Communications Across National Borders in Terrorist and Criminal Investigations*. Mont-Tremblant, Quebec, Group of 8, 2002d.

G8 Lyon Group 'Un dialogue entre les pouvoirs publics et le secteur privé sur la sécurité et la confiance dans le cyberespace', Communiqué du G8 (Groupe de Lyon) (Paris: G8, 2000).

G8 Lyon Group *Recommendations for Tracing Networked Communications Across National Borders in Terrorist and Criminal Investigations* (Draft), G8, 2001.

Gaspar, R. *Looking to the Future: Clarity on Communications Data Retention Law*. A submission to the Home Office for Legislation on Data Retention, On behalf of ACPO and ACPO(S); HM Customs & Excise; Security Service; Secret Intelligence Service; and GCHQ, 2000.

Goldsmith, J.L. 'Against Cyberanarchy', *University of Chicago Law Review*, 65 (1998), 1199–1250.

Goldsmith, J.L. 'Symposium on the Internet and Legal Theory: Regulation of the Internet: Three Persistent Fallacies', *Chicago-Kent Law Review*, 73 (1998) 1119–1131.

Johnson, D.R. and D.G. Post 'Law And Borders – The Rise of Law in Cyberspace', *Stanford Law Review*, 48 (1996) 1369.

Lugar, Serator R.G., Chairman. *Opening Statement for Hearing on Law Enforcement Treaties*, 17 June 2004. Available at http://foreign.serate.gov/testimony/2004/LugarStatement040617.pdf.

Purdy, D. *Report of the G8 Government/Private Sector High Level Meeting on High-tech Crime* (Tokyo, Japan: Department of Foreign Affairs and International Trade, 2001).

Sun, J.-M. and J. Pelkmans 'Regulatory Competition in the Single Market', in C. Hood (ed.), *A Reader on Regulation* (Oxford: Oxford University Press, 1998), pp. 443–67.

United States and Japan (2003) *United States–Japan Joint Statement on Promoting Global Cyber Security*.

United States Government *Comments of the United States Government on the European Commission Communication on Combating Computer Crime* (Brussels, 2001).

15
Conclusion

Dr Steve Wright, Dr Philippa Trevorrow, Professor David Webb and Dr Edward Halpin

What constitutes the greatest cyber threat in the uncertain years which lie ahead? The popular vision heavily promoted by the media is a terrorist fanatic group or a lone 'geeky whiz kid' with a genius IQ, actively seeking to bring down western civilization through hacking into sensitive locations such as NATO (North Atlantic Treaty Organization) high command or the US Air Traffic Control systems. The recent case of Gary McKinnon[1] highlights this popular stereotype. Designated by US prosecutors as the 'biggest military computer hack of all time',[2] McKinnon, at the time of writing, faces up to 70 years in jail for hacking into US government agencies, including US Space Command and DARPA (Defence Advanced Research Projects Agency).[3]

Is this the most realistic threat? Other commentators highlight the more hidden military research dedicated to attacking urban infrastructures or the massive effort by some states to control or monitor Internet traffic, whether in the European Union (EU) for commercial reasons, or in China for tracking potential political dissidents.

As mentioned in the Introduction to this volume, in the age of Information Warfare (IW) it should come as no surprise that the military see the Internet as just another tool and that many individuals, groups and organizations view the dependence of modern society on telecommunications infrastructure as a vulnerability to be exploited.

The advent of nano-technologies[4] will inevitably change the way in which such weapons and data are put together to achieve more effective target acquisition and destruction. 'Super miniaturization' will enable individual soldiers to become part of a more efficient battlefield where commanders use surveillance actually to see through the helmets of their men. Individual tagging will identify friend from foe and prevent losses through 'friendly fire' but, increasingly, surveillance technology will change the nature of targets in information-dependent societies.

Indeed, IW is part of a variety of new forms of emergent attack strategies. Most modern states are dependent upon telecommunications infrastructure and many within military circles are asking, 'Why destroy civilian infrastructure if a country's nervous systems can be disabled instead?' According to General Fogleman, the US Air Force Chief of Staff, 'Dominating the information spectrum is as critical to conflict now as occupying the land or controlling the air has been in the past.'[5]

This aspect of war fighting is rapidly changing, despite the lack of media attention. Some academics, such as Professor Stephen Graham of the University of Durham, have started to examine the largely unreported military efforts being designed to target urban infrastructure. The phenomena of 'unblackboxing' urban infrastructure has already become part of modern warfare tactics[6] and Graham has identified a mentality in many states to 'permanently demodernize urbanized adversaries'. Quoting *New York Times* columnist Thomas Friedman, he links this form of warfare with US hegemony whilst noting that China has similar war doctrines.

It should be lights out in Belgrade: every power grid, water pipe, bridge, road and war related factory has to be targeted...we will set your country back by pulverizing you. You want 1950? We can do 1950. You want 1389? We can do that too.[7]

Graham's analysis covers both first-order effects – the destruction of lights and refrigeration in buildings – and second-order effects – hygiene problems, shortage of drinking water, the inability to prepare and process some foods, together with tertiary effects which may involve an increased number of non-combatants requiring assistance. Although publicly presented as non-lethal denial of resources to the enemy, this type of technology viewed holistically dissects adversary societies. The consequences are summed up as 'bomb now, die later'.[8]

Modern surveillance technology is becoming part of that infrastructure. Weapons now have built-in primitive surveillance algorithms, but supposedly neutral telecommunications infrastructure such as the mobile phone network can be used not only for surveillance but also for pinpointing and targeting specific individuals and groups. Since targets are identified by coordinates, such digital location data carried around by most of us (e.g. through mobile phones or ID smart cards) can be used to programme target selection by weapons systems.

A primary goal of this book was to explore how the RMA is creating new technologies both to facilitate and to target surveillance infrastructure and how 'no hiding place' military doctrines will begin to infiltrate future urban living spaces. The dictates of a growing international crisis will also drive a move away from simply mass supervision to more exact systems of pinpoint targeting. We should, for example, be particularly interested in how some types of new border control technologies can incorporate punishment, with surveillance systems managed through the Internet to become victim-activated networks.

A recent presentation from the College of Aerospace, Doctrine Research and Education in the US,[9] for example, lists quantum computers, intelligent software, virtual reality, intelligent materials, directed-energy weapons, lasers, biotechnology, human/computer interfaces, mind control, micro-technology, millimetre wave cameras and video insertion amongst its emergent technologies. Many of these technologies have a surveillance dimension, whether it is present for targeting, for directed control, or for feedback on effectiveness.

Such technologies are already beyond the prototype stage and represent substantial commercial investments. Commercial companies such as the German military conglomerate Diehl Microsystems have reported manufacturing a range of what they call C4I (Command, Control, Communications, Computers and Intelligence) systems using high-powered radio frequency weapons. These include single-shot munitions to wipe out weapon electronics; repetitive systems which can stop other munitions working; and missiles which can destroy telecommunications in a battlefield for up to 3,000 square metres and can be used to target air defence sites, command centres CCD cameras and power supplies.

Diehl sees roles for its technology in protecting ships in harbours, suppressing electronics in terrorist strongholds and even mine destruction. Advanced hand-held suitcase versions can wipe out surveillance cameras, mobile phones, industrial controllers, television, monitors, radios, telephones, computers and even car electronics for up to a kilometre.[10] Fantastic capabilities for hostage rescue perhaps – but what potential nightmares will such a capability yield if adopted and adapted by terrorist groups?

Future researchers should be particularly interested in the military's use of surveillance, information and computer technology for internal control, counterrevolutionary and anti-terrorist operations. Yet, in many senses the variety of possible themes are potentially too vast – from mechanical roboflies[11] to act as military micro spies to the Echelon global telecommunications surveillance operated by the National Airspace System (NAS).[12]

Whilst Echelon is presented as a necessary evil in fighting international crime and in prosecuting 'the war against terror', the key players in this hugely expensive international surveillance game were challenged by the European Parliament in the late 1990s with having their own economic and political control agenda.[13]

The USA's Echelon project failed to halt September 11 and yet the agencies involved, such as the NSA, have been rewarded with increased expenditure and a more invasive political brief. The telecommunications infrastructure run by the NSA also has a military brief as it plays an important role in the burgeoning missile defence programme. Very little work has been accomplished on examining how such global military surveillance networks affect civil society. We know that after September 11 other agencies will be programmed with the intelligence of such data but, without adequate independent supervision, this apparatus is capable of transcending any protections in national law which may exclude the mass surveillance of private life. For example, formerly telephone tapping was only permitted by warrant and not mass unauthorized trawl, which could now be the case.

As the international crisis deepens, groups which question the role and function of such agencies will in fact move up the potential food chain of targets of interest. This conclusion poses a simple question – 'How will we research such mega surveillance entities in times to come?' A few individuals and NGOs (nongovernmental organizations) such as Statewatch have done a sterling job in fitting together the topography of such transnational initiatives from open sources.[14] In a time of terror to what extent will future researchers have the freedom to investigate state structures and how can the readers of books, such as this, defend that activity? All else is merely social astronomy (observation without responsibility).

Notes

1. 'Game Over', *The Guardian*, 9 July 2005.
2. Ibid.
3. See interview with J. Ronson, 'Game Over', *The Guardian Weekend*, 9 July 2005, pp. 26–31.
4. J. Altmann, 'Military Uses of Microsystem Technologies', *Series Science, Disarmament and International Security* (Germany: FOANS 2).
5. General Fogelman, 'Information Operations', *US Air Force Doctrine Document 2-5*, 5 August (1998). Available at http://www.dtic.mil/doctrine/jel/service_pubs/afd2_5.pdf.
6. See S. Graham, 'Switching Cities Off', *City*, 9(2) (July 2005), 169–93.
7. New York Times columnist T. Friedman, 23 April 1999, cited by Graham, 2005.
8. 'Bomb Now, Die Later', *Washington Post*, (1998), p. 1. Available online at http://www.washingtonpost.com/wp-srv/inatl/longterm/fogofwar/vignettes/v10.htm.
9. W.A. Stanmeyer, *Emerging Technologies IW-270* (College of Aerospace: Doctrine Research & Education). Available online at http://www.afrl.af.mil.
10. See R. Stark, M. Sporer, G. Staines (DIEHL Munitionssysteme), 'Non-Lethal Capabilities: Facing Emerging Threats', *Compact High Power RF Sources For Non-Lethal Applications*, presentation, 2nd European Symposium on Non-Lethal Weapons, 13–14 May 2003, Ettlingen, Germany, section 19–19.
11. 'Mechanical "Roboflies" Lend Wings to Defence', *Financial Times*, 22 November 2001, p. 15. See also http://www.newswise.com/articles/view/502903/.
12. See http://www.jya.com/stoa-atpc.htm.
13. D. Campbell, *Interception Capabilities* (2000). Available online at http://www.iptvreports.mcmail.com/stoa_cover.htm.
14. See: http://www.statewatch.org.

Index